CYTOGENETICS

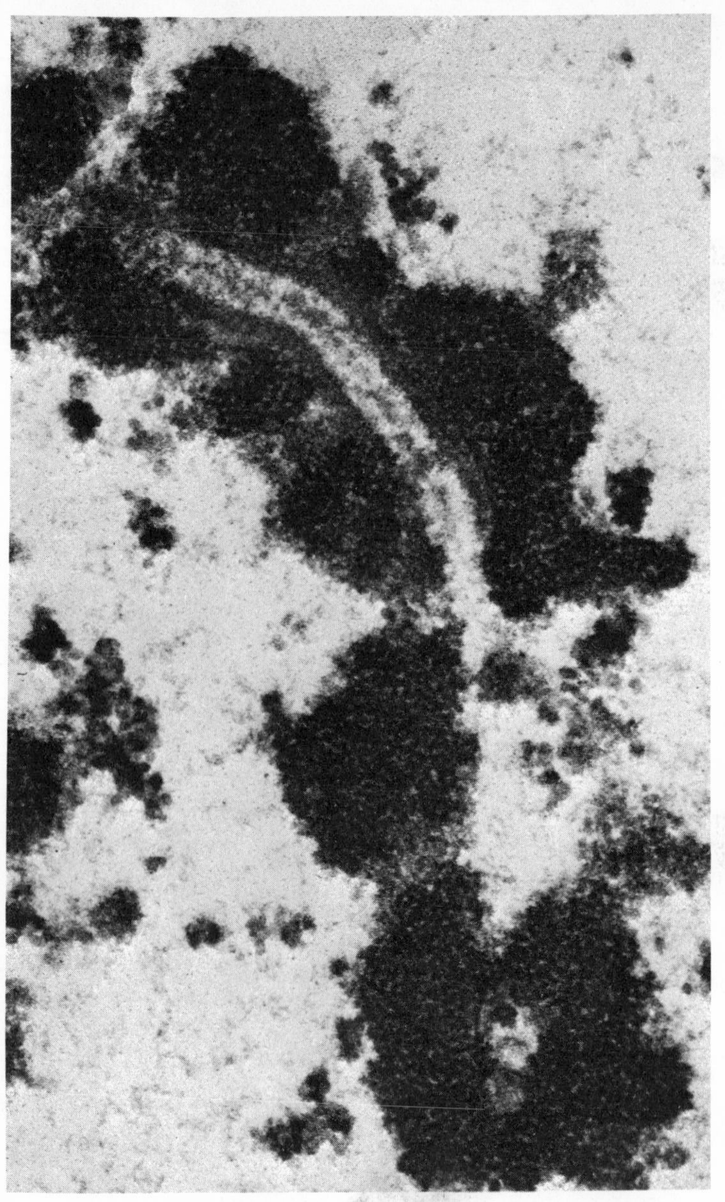

Electron micrograph of a synaptinemal complex at midpachytene in a pollen mother cell of Lilium. (Photograph courtesy of Dr. T. F. Roth.)

CONTENTS

PREFACE

Many courses in genetics include an exposure to Mendelian principles including linkage, the nuclear events during mitosis, meiosis, and fertilization, and chromosomal aberrations to support the laboratory exercises en route to a comprehensive treatment of genetic fine structure, DNA, protein synthesis, and enzyme regulation. Students with a vested interest in the genetics of plants or animals usually do not have access to a course in cytogenetics. Although this book was written with these students in mind, it also serves as a basic source for such a course.

It has been possible to consider the cytogenetic consequences of changes in the content, morphology, or number of standard chromosomes in almost Euclidean terms. Standard chromosomes are assumed to have one centromere, nonhomologous arms, and a characteristic linear sequence of Mendelian genes. Furthermore, crossing-over involves the mutual exchange of segments between homologous chromosomes or chromosomal segments according to the Bridges model. The basic principles of cytogenetics provide the axioms and theorems. The unusual chromosomes, aberrant situations, or nonconforming species furnish the exceptions which illuminate the generalizations.

The successful application of cytogenetic principles to the interpretation of genetic data from microbial species is a tribute to the cytogeneticists who accumulated the evidence for these principles during the long years in hot fields and humid greenhouses, in rooms filled with teeming pint bottles, or among endless cages.

I wish to thank those students at the University of Chicago whose vociferously positive and negative responses to the course in cytogenetics helped shape the content and organization of this book, Dick Gethmann, who examined the manuscript with the jaundiced eye of the competent graduate student, and Dr. Charles M. Rick, whose perceptive comments, criticisms, and suggestions played a significant role in the completion of the manuscript. I am indebted to Prof. I. Spergel for coffee and spiritual support. I have mixed feelings about Jim Young, of McGraw-Hill, who planted the notion that an introductory text on cytogenetics was needed. Now we shall have to endure a love-hate relation for years to come. How do I thank Rosalie who edited and prepared the manuscript in the eye of hurricanes Joeli, Martha, and Janie?

I am indebted to good friends and colleagues who searched their files for the original photographs in the book. Although their contributions are recognized in the appropriate legends, I want to thank them for their kindness and help.

E. D. Garber

1 INTRODUCTION

Schleiden and Schwann (1838–1839) focused attention on the cell as the basic organizational unit of life and thereby initiated the new biology. In time, the light microscopists were able to describe the basic phenomena of cellular and organismal reproduction in terms of the two organelles directly related to heredity: the nucleus and the chromosome. Once the sequence of events during mitosis, meiosis, and fertilization was established, the cell could also be viewed as the unit of heredity.

Three basic concepts associated with cellular and organismal heredity were formulated by the light microscopists: (1) Cells arise from preexisting cells. The rejection of the theory of spontaneous generation at the microbial level ended all opposition to the acceptance of the continuity of life forms. (2) Nuclei arise from preexisting nuclei. While cell (cytokinesis) and nuclear (karyokinesis) division is usually correlated in time, nuclear division is clearly the primary event. Furthermore, the essential feature of fertilization is the pooling of the contents of gametic nuclei to produce a zygotic nucleus. (3) Chromosomes arise from preexisting chromosomes. Although the constancy of the number and morphology of the chromosomes in successive mitotic divisions or in the individuals of a species supports this concept, the conclusive evidence comes from cytogenetics.

Cytogenetics convincingly demonstrated that the chromosomes are the vehicles for the genes without proposing an acceptable model to account for the physicochemical organization of the chromosome or the gene. Microbial genetics, biochemistry, and biophysics furnished the evidence to construct such models, which also attempted to include the observations provided by the electron microscopists. A fourth concept of heredity states that genetic information arises from preexisting genetic information, thereby unifying prokaryotic and eukaryotic genetics. The prokaryotes include viruses, bacteria, and blue-green algae, and the eukaryotes, the other life forms.

The significant advance in understanding the physicochemical organization of the chromosome came from genetic studies with bacteria and bacterial viruses and not from the cytogenetics of eukaryotes. While transformation in pneumococcus was shown to involve deoxyribonucleic acid (DNA), the elegant experiments of Hershey and Chase (1952) clearly demonstrated that the genetic information of T4 coliphage was associated with the DNA and not the protein. With the double helix of polynucleotides proposed by Watson and Crick (1953) to describe the prokaryotic chromosome, it was possible to view the eukaryotic chromosome from new angles and to approach old questions with fresh resources.

Molecular biology and electron microscopy were responsible for establishing the basic taxonomic dichotomy for the life forms: prokaryote or eukaryote. In prokaryotes, the genetic information is contained in a single circular chromosome identifiable as a double helix of polynucleotides without associated proteins and not enclosed in a membrane. In some viruses, the chromosome is a single DNA or RNA strand. In the eukaryotes, the genetic information is distributed in two or more linear chromosomes of complex chemical and physical organization but with at least one double helix, and all the chromosomes are enclosed in a membrane when the nucleus is not dividing. Furthermore, the elaborate processes of mitosis and meiosis for distributing the genetic information from one eukaryotic nucleus to another are not found in the prokaryotes.

From a genetic point of view, sex represents the means for getting new combinations of genes. Breeding experiments with eukaryotes correctly assumed that the fusion of nuclei is necessary to produce an aggregate (the zygotic nucleus) from which new gene combinations can eventually emerge. Similar experiments with prokaryotes, however, indicated that a less restrictive definition would be required to unify the genetic interpretations of breeding data from both prokaryotes and eukaryotes. Sex can be defined as the pooling of genetic information from different sources for eventual distribution to progeny. With the localization of the genetic material and information in the double helix of both prokaryotic and eukaryotic chromosomes, the basic questions concerning the physicochemical organization and replication of the eukaryotic chromosome are currently being referred to the organization and replication of the prokaryotic chromosome. Unfortunately, so far such ventures have neither resolved these questions nor provided acceptable molecular interpretations for the cytogenetic data from eukaryotes.

CHROMOSOME STRANDEDNESS

During the period when the stages of mitosis and meiosis were being described in terms of the nuclear events observed with the light microscope, the chromosomes received considerable attention. In mitosis, the prophase and metaphase chromosomes appeared to be two-stranded and the anaphase chromosomes, one-stranded. Consequently, the replication of the strands or chromatids presumably occurred during telophase or interphase. In meiosis, the leptotene and zygotene chromosomes seemed to be one-stranded, and, to account for the obviously two-stranded diplotene chromosome, chromatid replication was assumed to have occurred during the intervening pachytene. By 1916, Bridges obtained genetic evidence from an investigation of crossing-over between sex-linked genes in *Drosophila melanogaster* which was interpreted by assuming that crossing-over involved chromosomes with two chromatids. In 1931, Darlington proposed the "precocity theory" to integrate the contemporary cytological and genetic observations by invoking one- or two-stranded chromosomes.

In the course of describing the metaphase or pachytene chromosomes of plant and animal species, cytologists discovered that certain topological features could be used to characterize the chromosomes. Each chromosome has a specialized segment, the *centromere*, which becomes associated with fibers of the spindle apparatus. The centromere of each chromosome occupies a constant site so that each chromosome can be *metacentric*, with arms of equal length; *submetacentric*, with arms of clearly different lengths; or *acrocentric*, with one extremely short and one extremely long arm. Usually, only one chromosome is responsible for the appearance of the nucleolus at a specialized segment occupying a particular site on the chromosome. In a number of species, one or more chromosomes have a terminal segment, the *satellite*, which is separated from the rest of the chromosome by a "stalk." Finally, the pattern of small or large darkly staining bodies, chromomeres and knobs, respectively, along the pachytene chromosomes and the presence of darkly staining segments at constant sites in the pachytene chromosomes are useful morphological landmarks in characterizing each chromosome arm as well as the different chromosomes in a species. Once the cytogeneticists became involved with the placing of genes in specific regions of the chromosomes, the exquisitely detailed salivary-gland chromosomes of *Drosophila* were probably responsible for the great interest in the cytogenetics of this genus.

All the models for the physical organization of DNA in

eukaryotic chromosomes can be reduced to a one-stranded (unineme), two-stranded (bineme), or multistranded (polyneme) structure. The unineme chromosome with one chromatid containing a single double helix is assumed to yield two chromatids, each with a single double helix. The bineme or polyneme chromosome has subchromatids (two strands) or quarter chromatids (four strands) within the chromatid or more than one double helix in each chromatid. For some time the unineme chromosome was the accepted model, but sufficient evidence for bineme or polyneme chromosomes has now accumulated to challenge that model. Increasing the numbers of strands within the eukaryotic chromosome complicates the interpretations of most cytogenetic data, which may account for the reluctance of cytogeneticists to consider polyneme chromosomes as appropriate models.

Occasional reports of multistranded anaphase chromosomes by light microscopists were not seriously considered and were assessed as mere artifacts. Recent studies by light microscopists have indicated multistranded metaphase (Fig. 1-1) and anaphase (Fig. 1-2) chromosomes, and these observations cannot be lightly disregarded. The importance of the number of strands in the eukaryotic chromosome is responsible for the numerous efforts to resolve this problem and to relate chromosome strandedness to cytogenetic data.

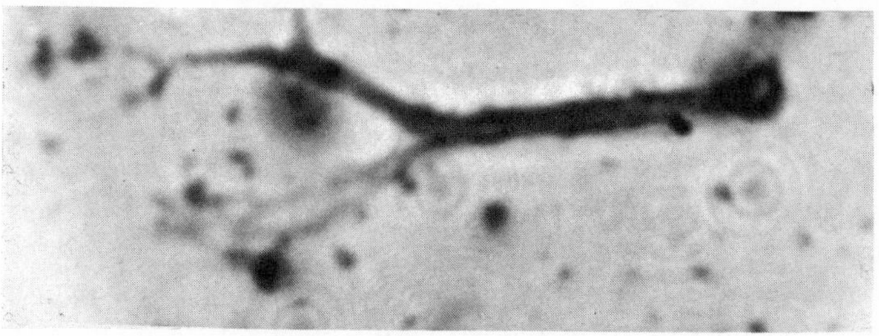

Fig. 1–1. Feulgen-stained, mildly trypsin-treated metaphase chromosome of *Vicia faba*, showing one chromatid separated into half-chromatids, with each of these bifurcating into quarter chromatids. (*Photograph courtesy of Dr. S. Wolff; Wolff, 1965.*)

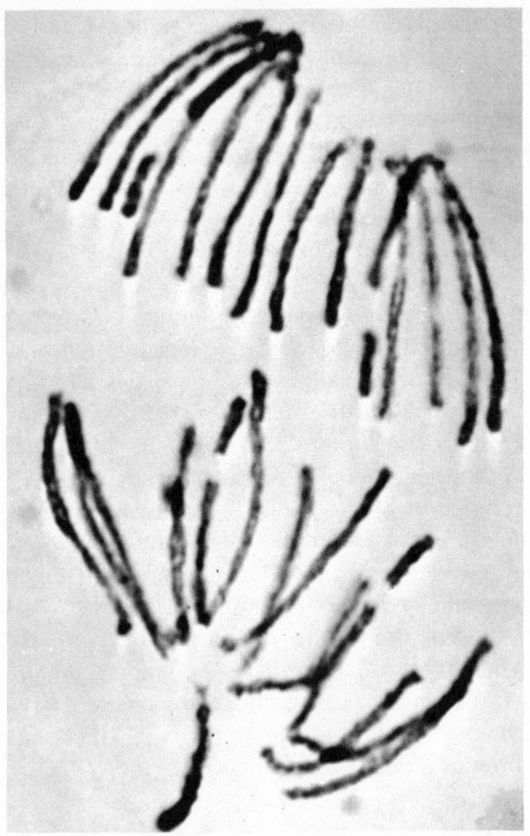

Fig. 1–2. Relationally coiled half chromatids of the anaphase chromosomes of *Vicia faba*. (*Photograph courtesy of Dr. S. Wolff; Peacock, 1965.*)

Cells exposed to ionizing radiation or chemical agents usually give chromosomal configurations at anaphase which can be interpreted as products of the fusion of broken strands in either one- or two-stranded chromosomes. In a number of cases, the separation of sister chromatids during mitotic or meiotic anaphase may be impeded by a side-arm bridge (Fig. 1-3 *a* and *b*). To produce such a bridge, each chromatid must

Diplotene bivalent	Anaphase I	Description
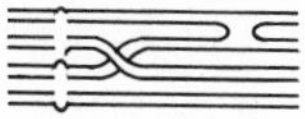		Two side–arm bridge (true)
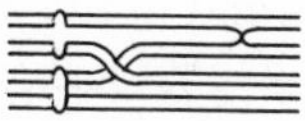		Two side–arm bridge (pseudo)
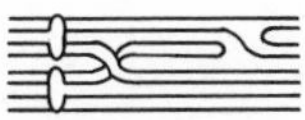		Y chromosome and deficient chromatid
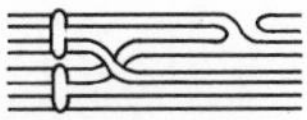		One side–arm bridge and fragment

Fig. 1–3. Anaphase I configurations resulting from some of the patterns of the reunion of half chromatids following a break across a diplotene chromosome. The anaphase I configurations result from a crossover proximal to the breakage-reunion site; second-meiotic-division configurations occur in the absence of the crossover. (*Peacock, 1965; after Wilson et al., 1959.*)

have at least two half chromatids or subchromatids. While such observations indicate multistranded chromosomes, John and Lewis (1965) take a conservative position: "In conclusion, while the evidence for multistrandedness, at least in some species, is accumulating, that for subchromatid exchange is not compelling except in a few cases."

Lampbrush chromosomes are found as bivalents in the oocytes of amphibians with large eggs and considerable yolk. At diplotene, the bivalents are extremely long, up to 800 microns, and resemble a brush with long symmetrical loops emerging from the longitudinal axis (Fig. 1-4). The loops originate from chromomeres along the axis, and while opposite loops are the same length, they are not the same length along the axis of paired chromosomes. Gall (1963) followed the kinetics of the breakage of *Triturus* lampbrush chromosomes by deoxyribonuclease to determine their strandedness. The number of breaks in the longitudinal axis and in the loops was determined and plotted against time by photographing the chromosomes at intervals during the exposure to the enzyme. Assuming that each chromosomal subunit was inde-

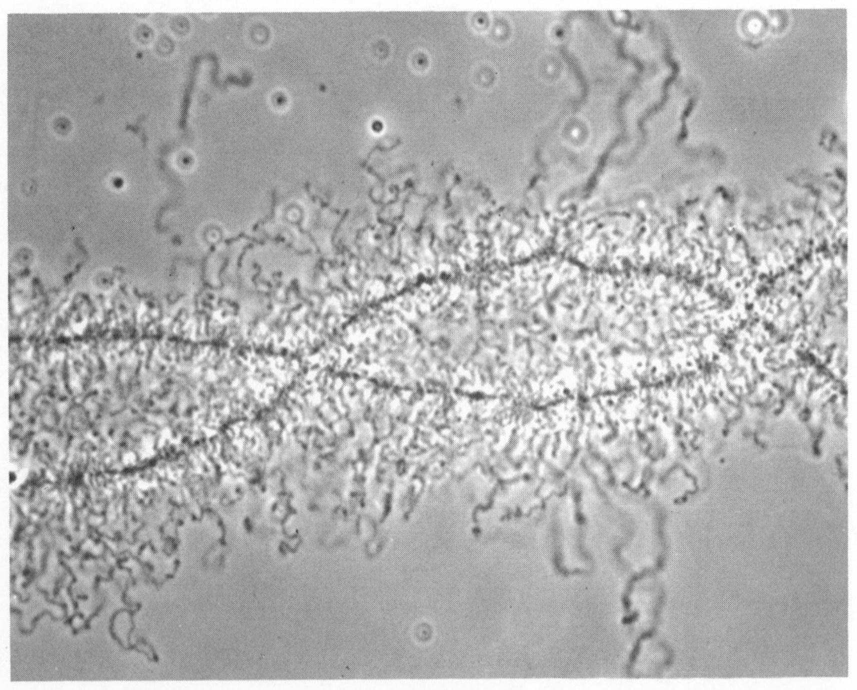

Fig. 1–4. Phase photomicrograph of part of a lampbrush chromosome of *Triturus viridescens* showing two chiasmata and the lateral loops from the axial strands. (*Photography courtesy of Dr. J. G. Gall; in Taylor, 1963.*)

pendently and equally likely to be cleaved by the enzyme, the number of breaks should be directly proportional to the time for one subunit, proportional to the square of the time for two subunits, and so forth. The calculated values from the data were 2.6 ± 0.2 subunits for the loops and 4.6 ± 0.4 subunits for the longitudinal axis. These values agreed with the visual observations, which indicated that the loops are parts of individual chromatids and that the axis is a pair of sister chromatids. Furthermore, the results are compatible with the assumption that each chromatid contains two subunits, the polynucleotides of the double helix.

Electron microscopy was expected to provide significant information concerning the strandedness of eukaryotic chromosomes and had clearly demonstrated the circular chromosomes of prokaryotes, which were readily interpreted as the polynucleotides in the double helix. The eukaryotic chromosome in resting nuclei or in nuclear division has been examined in whole mounts and in sections by electron microscopy and found to contain highly folded and closely packed fibers (Fig. 1-5). The fundamental fiber, or unit, is approximately 100 Å in diameter, and each unit appears to include a duplex of fibrils, approximately 40 Å in diameter. The temptation to interpret the duplex fibrils as the double helix of polynucleotides has not been successfully resisted. Cytogeneticists had anticipated a more meaningful physical organization of the genetic material in the eukaryotic chromosome. Furthermore, electron micrographs have not yet provided any clues to the physical relation between the polynucleotides and the proteins present in the eukaryotic chromosomes.

CHROMOSOME REPLICATION

Once the genetic material in eukaryotic chromosomes had been directly related to DNA, chromosome replication could be investigated in terms of DNA synthesis. These investigations were expected to furnish evidence which would establish

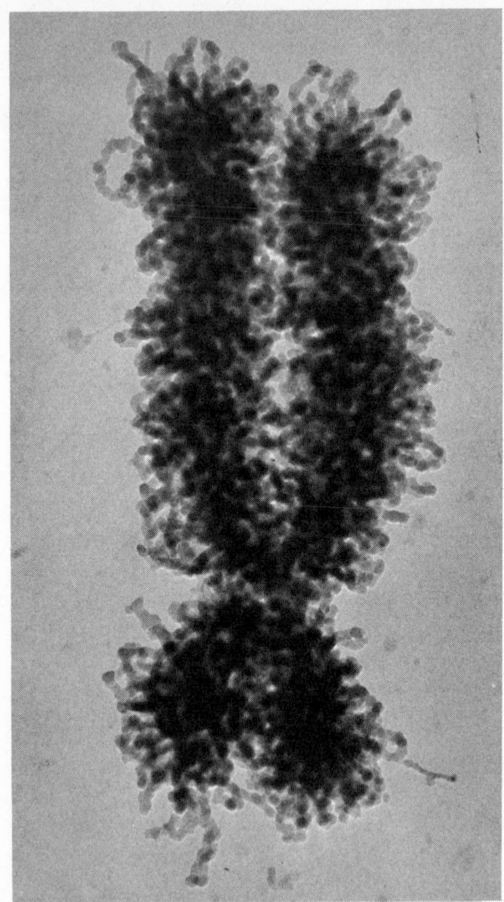

Fig. 1–5. Electron micrographs of a whole mount of human chromosome 12, showing the centromere and sister chromatids. Note the looping of the strands at the ends of the chromatids. (*Photograph courtesy of Dr. E. J. DuPraw. From E. J. DuPraw. 1970. DNA and chromosomes. Holt, Rinehart and Winston, Inc., New York.*)

a basis for determining the strandedness of chromosomes. Cytospectrophotometric measurements and quantitative analysis showed a constant relation between the amount of DNA and the number of chromosome sets in the nucleus. In dividing cells, the quantity of DNA per nucleus doubled during interphase and was equally distributed between the two groups of chromosomes at anaphase or telophase. To determine the precise timing of DNA synthesis during interphase, isotopically labeled compounds which were incorporated into the DNA of replicating chromosomes were detected by autoradiography.

Howard and Pelc (1953) immersed roots of *Vicia faba* in a solution containing P^{32} and removed samples of root tips at intervals to determine when the isotopic label was incorporated into chromosomal DNA. The period of DNA synthesis (S) was estimated at approximately 6 of the 26 hr of interphase. The interval, or gap (G_1), between the preceding late telophase and the S period was estimated to be 12 hr; the gap (G_2) between the S period and early prophase of the next division was estimated to be 8 hr. Similar experiments using a number of plant and animal cells have been interpreted in terms of the G_1, S, and G_2 periods, which can have different values, depending on the material. The timing of DNA synthesis in the interphase nucleus, the amount of DNA per nucleus, and the chromosome number have been used to construct models for the strandedness of chromosomes (Fig. 1-6). The same information, however, will support a one-stranded or a two-stranded (subchromatids) chromosome.

The chromosomal configurations at anaphase after cells have been exposed to ionizing radiation during interphase can be interpreted using the knowledge of the period in which the chromosomes are irradiated. Cells treated during the G_1 period gave configurations for breaks and fusions in one-stranded chromosomes which later replicated to produce two-stranded chromosomes; cells irradiated at the S or G_2 period gave configurations compatible with breaks and fusions for a two-stranded chromosome. As already mentioned, the detec-

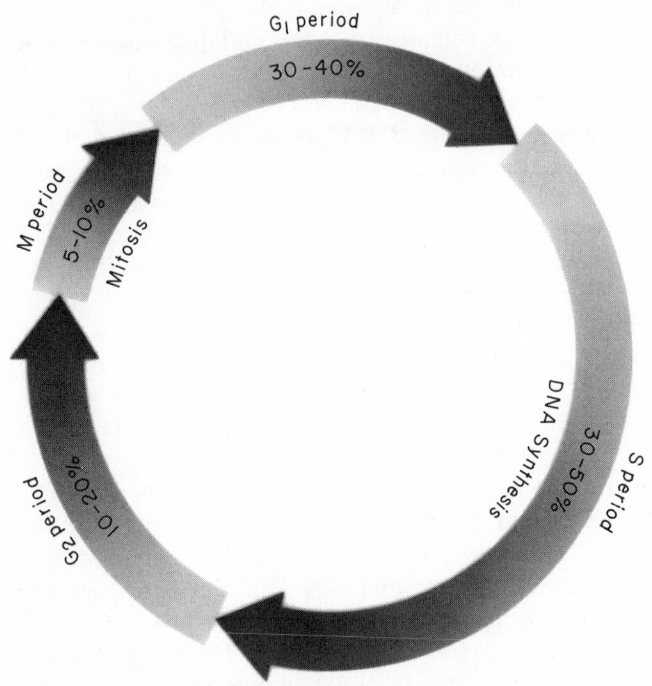

Fig. 1–6. DNA synthesis during the mitotic cycle. The G_1 (gap) period extends from the previous telophase to the beginning of the S (synthesis) period. The S period is characterized by the doubling of the DNA content from 2C to 4C. The G_2 period is the interval between the end of DNA synthesis and the onset of the next mitotic (M) period.

tion of configurations with side-arm bridges at anaphase calls for two subchromatids in each chromatid.

The double helix of polynucleotides may replicate conservatively or semiconservatively; that is, one double helix serves as a template for the production of another double helix, or each strand of the double helix separates and serves as a template for the production of the complementary strand. These possibilities were investigated in the expectation that a decision might lead to an understanding of the strandedness of chromosomes. Taylor, Woods, and Hughes (1957) labeled the root-tip chromosomes of *V. faba* with tritiated thymidine, which was specifically incorporated into the replicating DNA. The labeled chromosomes were followed through successive mitotic divisions after colchicine treatment by autoradiography. During metaphase, the chromatids of each chromosome diverged so that the presence or absence of the label in each chromatid was readily detected. Furthermore, colchicine destroyed the spindle apparatus in each of the mitotic divisions without interfering with chromosome replication and separation. Consequently, the number of mitotic divisions after labeling could be determined by counting the number of chromosomes. After treatment with tritiated thymidine, both chromatids of each chromosome were uniformly labeled. The first mitotic division after labeling gave chromosomes with one unlabeled and one labeled chromatid; the second mitotic division after labeling yielded chromosomes with two unlabeled chromatids or with one labeled and one unlabeled chromatid. The observations were interpreted by assuming a semiconservative replication of two units (strands) in the chromosome (Fig. 1-7): "The chromosomes before duplication are composed of two units which extend throughout the length of the chromosomes. The units separate at duplication and each has a complementary unit built along it."

The occasional appearance of two labeled chromatids in chromosomes in the second mitotic division after labeling raised questions concerning the relation between DNA replication and the strandedness of chromosomes. The presence of isolabeled chromatids has stimulated the construction of models for chromosomes in which the chromatids have at least two double helixes and in which the distribution of the label will yield not only the usual observations but also account for the isolabeled chromatids.

Meselson and Stahl (1958) investigated the replication of DNA in the prokaryotic chromosome of one double helix by applying the same rationale used to label the DNA of eukaryotic chromosomes with tritiated thymidine. Cells of

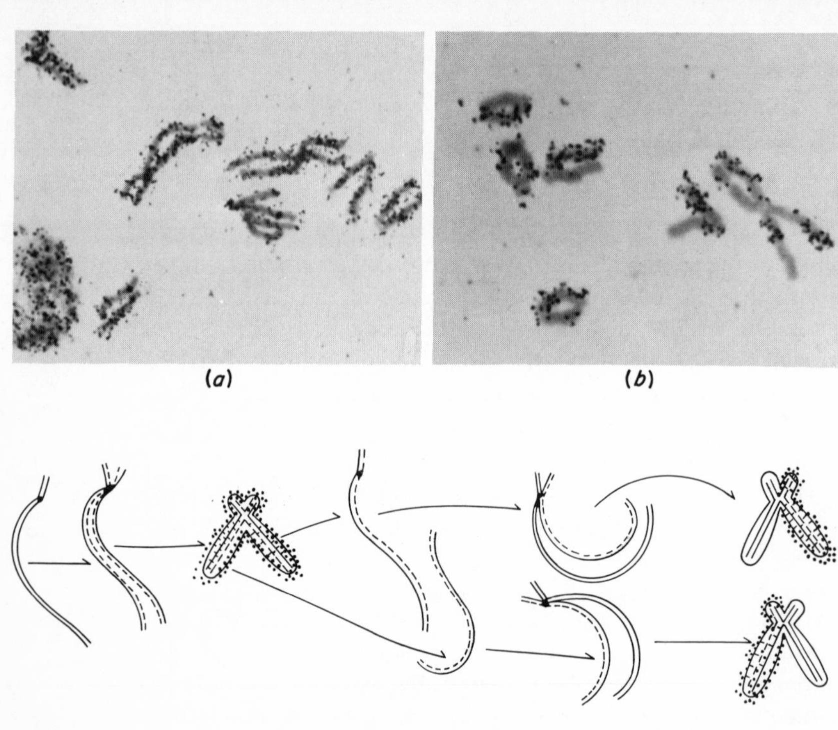

Fig. 1–7. Replication of DNA in the somatic chromosomes of *Vicia faba*. (*a*) Autoradiograph of the chromosomes after one replication cycle in which tritiated thymidine was incorporated into the DNA of each chromatid. (*b*) Autoradiograph of the part of the chromosome complement of one root cell which had incorporated the label during one replication cycle and then replicated once without added label. The labeled DNA subunits have segregated into separate chromatids, which also have been involved in sister-chromatid exchange. (*Photographs courtesy of Dr. J. H. Taylor; Taylor, 1962.*) (*c*) Distribution of subunits to explain the segregation of the labeled chromatids in the photographs. The broken lines represent labeled subunits, and the unbroken lines, unlabeled subunits. (*Taylor et al., 1957.*)

Escherichia coli were grown in a defined medium containing ammonium chloride as the sole source of nitrogen, and the extracted DNA was subjected to high-speed centrifugation in a cesium chloride density gradient until an equilibrium was reached for the site of the DNA. After growth in a medium containing the heavy isotope N^{15}, the extracted DNA had a significantly different site in the density gradient. Cells grown in the N^{15} medium were shifted to the N^{14} medium; samples of cells were withdrawn from the culture at specific intervals during each of successive generations; and the extracted DNA was subjected to density-gradient centrifugation to determine the site of equilibrium. After one generation, all the DNA occupied a site at a position midway between the sites for the N^{15}- or N^{14}-labeled DNA. After the second generation, two sites were detected, one at the midway position and the other at the N^{14} position. These observations were consistent with a semiconservative replication of the polynucleotide strands of the double helix in prokaryotic chromosomes. Sueoka (1960) employed density-gradient centrifugation to obtain evidence for the semiconservative replication of the DNA in the chromosomes of the eukaryotic alga *Chlamydomonas*.

The physicochemical organization of the eukaryotic chromosome is clearly more complex than that of the prokaryotic chromosome. One approach to investigating the strandedness of each type has considered the recombination of linked genes or crossing-over between homologous chromosomal segments.

Bridges (1916) explained unexpected breeding data from an investigation of sex-linked genes in XXY females of *D. melanogaster* by assuming that crossing-over involves the mutual exchange of segments between nonsister chromatids of homologous chromosomes. Extensive cytogenetic studies on crossing-over indicated a high degree of precision for this mutual exchange of chromatid segments. Although it seemed reasonable to diagram the results of crossing-over in terms of the coordinated breaks and fusions for nonsister chromatids, no mechanisms were offered to account for the breakage and fusion of chromatids or for the precision of breakage at the same sites in nonsister chromatids. Belling (1928) proposed a prescient model which considered the replicating chromatids as the source of the crossing-over, but the model did not explain three-strand or four-strand double crossing-over without appealing to the breakage and fusion of the two template chromatids.

The discovery of "sex" in prokaryotes seemed to offer new approaches in understanding the nature of crossing-over. Breeding data were manipulated to yield a single, circular linkage map; it was eventually confirmed by electron micrographs which showed a circular chromosome. In transformation and transduction, a segment of genetic information (DNA) from a donor is integrated into the chromosome of the recipient. During conjugation in *E. coli,* the ring chromosome opens, and the linear chromosome enters the female cell, so that genetic recombination in this situation also involves a linear segment of genetic information. Finally, the genetic information of such episomes as the F factor and temperate bacteriophage is contained in a ring chromosome and can become incorporated into the ring chromosome of the bacterial cell. In all these situations involving the incorporation of genetic information by transformation or transduction, the recombination in *E. coli,* and the incorporation of genetic information of the episomes, it has been assumed that crossing-over has been responsible. Furthermore, crossing-over has been assumed to involve two-stranded (the double helix) chromosomes or chromosomal segments.

CROSSING-OVER

Attempts to interpret the recombination data from linkage studies in *E. coli* led to the revival of the Belling hypothesis in the form of the *copy-choice mechanism* for the replication of the prokaryotic chromosome (Lederberg, 1955). According to this mechanism, a recombinant chromosome is the result of a replication in which one parental and then the other parental strand is used as the template. Once the prokaryotic chromosome was characterized as a double helix which replicated semiconservatively, the copy-choice hypothesis was abandoned. Experiments with isotopically labeled DNA in the bacteriophage lambda indicated that recombinants resulted from the mutual exchange of segments between unreplicated polynucleotides, that is, breakage and fusion (Meselson, 1964).

Genetic studies with the ascomycetous fungi can be accomplished by the analysis of randomly collected ascospores or of the four or eight ascospores from each ascus. Although the expected ratio of 1 mutant:1 wild type is generally found for the ascospores from one ascus, deviations from this ratio have been reported for a number of species. In certain species, asci with eight ascospores have yielded ratios of 6:2 or 2:6, and in other species, asci with four ascospores, ratios of 1:3 or 3:1. This phenomenon has been termed *gene conversion.* Moreover, crossing-over usually occurs in the immediate region of the converted allele. These observations have been incorporated in a molecular (DNA) model to account both for gene conversion and for crossing-over (Whitehouse and Hastings, 1965). Although the model has been modified, the basic scheme has not been significantly altered.

According to the Whitehouse model (Fig. 1-8), chromatid replication is essentially complete prior to crossing-over. When the homologous chromosomes pair at zygotene or have completed pairing at pachytene, crossing-over can occur at several sites along the chromosomes and generally involves nonsister chromatids. Finally, crossing-over requires the breakage of one polynucleotide in each double helix of the chromatids and their reconstruction by the lateral association of complementary segments from homologous regions to produce recombinant polynucleotides with a short biparental segment, that is, derived from each parental polynucleotide strand. While a small amount of DNA synthesis is required to pro-

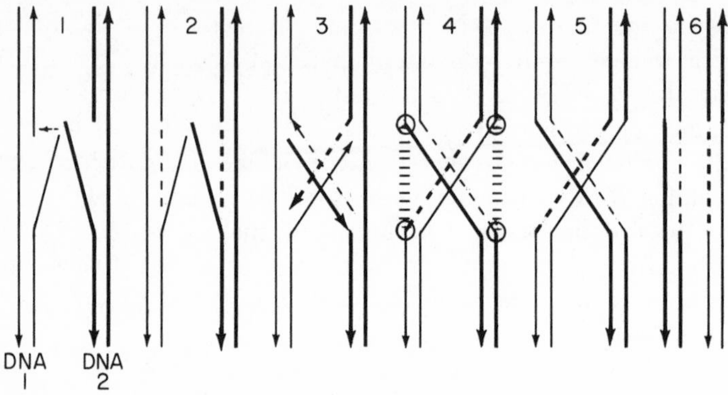

Fig. 1–8. A model to account for recombination by the breakage and reunion of DNA polynucleotides in different twin helixes. Each strand is indicated by a line, with its polarity shown by an arrow. The hydrogen bonds between strands are omitted. (*1*) Single-strand breaks occur in strands of opposite polarity (*arrows*) in the homologous helixes. The helical structures "melt" locally. (*2*) The broken single strands are extended by synthesis (*dashed lines*), copying the unbroken complementary strand. (*3*) The newly synthesized segments peel off to pair with those that had peeled off in step 1. (*4*) The remaining single strands are degraded hydrolytically (*hatching*), and the discontinuities (*circles*) are enzymatically closed. (*5*) Two recombinant molecules are formed, which are separately shown in the next diagram. (*6*) Each recombinant is heteroduplex: one strand from one parent (*heavy line*) and the other strand from the other parent (*thin line*). (*Davis et al., 1967; after Whitehouse and Hastings, 1965.*)

duce the recombinant segments, the total amount of DNA shows no net gain. Hotta, Ito, and Stern (1966) demonstrated a small amount (approximately 0.3 percent) of a seemingly specific pattern of DNA synthesis during zygotene and pachytene in the pollen mother cells of lily, and an inhibition of this synthesis by 5-fluorouridine produces a failure of the normal pairing of the meiotic chromosomes. These observations support the Whitehouse model. The small amount of DNA synthesis at certain sites can be viewed as the breakage and fusion of polynucleotides to produce the chiasmata. According to this model, interference with DNA synthesis would be expected to preclude the formation of chiasmata.

Molecular models of chromosomes to account for crossing-over in prokaryotes and eukaryotes will probably assume that each chromatid has at least one double helix. It is not yet clear how the models for the eukaryotic chromosome will view the proteins associated with the DNA. To account for high negative interference, polarity, and nonreciprocal exchange in intragenic recombination, chromatid segments may have to include more than one double helix. The necessary enzymes to accomplish the different steps involved in crossing-over have already been assigned names (endonuclease, DNA polymerase, DNA ligase), and enzymes with these properties have been isolated.

Although models of chromosomes will probably continue to be generated to meet the demands of new observations, such models will eventually have to consider chromosome pairing (synapsis). Meanwhile, cytogeneticists still get considerable mileage from Bridges' 1916 model in interpreting the breeding data and chromosomal configurations for eukaryotes with structurally altered chromosomes or aberrant chromosome number.

REFERENCES

BELLING, J. 1928. A working hypothesis for segmental interchange between homologous chromosomes in flowering plants. Univ. California Pub. Bot. 14:283–291.

BRIDGES, C. B. 1916. Nondisjunction as proof of the chromosome theory of heredity. Genetics 1:1–52, 107–163.

DARLINGTON, C. D. 1931. Meiosis. Biol. Rev. 6:221–264.

DAVIS, B. D., R. DULBECCO, H. N. EISEN, H. S. GINSBERG, AND W. B. WOOD. 1967. Microbiology. Hoeber Harper, New York.

DUPRAW, E. J. 1970. DNA and chromosomes. Holt, New York.

GALL, J. G. 1963. Kinetics of deoxyribonuclease action on chromosomes. Nature 198:36–38.

HERSHEY, A. D., AND M. CHASE. 1952. Independent functions of viral protein and nucleic acid in growth of bacteriophage. J. Gen. Physiol. 36:39–56.

HOTTA, Y., M. ITO, AND H. STERN. 1966. Synthesis of DNA during meiosis. Proc. Nat. Acad. Sci. 56:1184–1191.

HOWARD, A., AND S. PELC. 1953. Synthesis of desoxyribosenucleic acid in normal and irradiated cells and its relation to chromosome breakage. Heredity (Suppl.) 6:261–273.

JOHN, B., AND K. R. LEWIS. 1965. The meiotic system. Protoplasmatologia 6 (F/1): 1–335.

LEDERBERG, J. 1955. Recombination mechanisms in bacteria. J. Cell. Comp. Physiol. (Suppl.) 45:75–107.

MESELSON, M. 1964. On the mechanism of genetic recombination between DNA molecules. J. Mol. Biol. 9:734–745.

——— AND F. W. STAHL. 1958. The replication of DNA in *Escherichia coli*. Proc. Nat. Acad. Sci. 44:671–682.

PEACOCK, W. J. 1965. Chromosome replication. Nat. Cancer Inst. (U.S.A.) Monogr. 18, pp. 101–131.

SUEOKA, N. 1960. Mitotic replication of deoxyribonucleic acid in *Chlamydomonas reinhardi*. Proc. Nat. Acad. Sci. 46:83–91.

TAYLOR, J. H. 1962. Chromosome reproduction. Int. Rev. Cytol. 13:39–73.

———. (ed.). 1963. Molecular genetics. Academic, New York.

———, P. S. WOODS, AND W. L. HUGHES. 1957. The organization and duplication of chromosomes as revealed by autoradiographic studies using tritium-labeled thymidine. Proc. Nat. Acad. Sci. 48:122–128.

WATSON, J. D., AND F. H. C. CRICK. 1953. A structure for deoxyribose nucleic acid. Nature 171:737–738.

WHITEHOUSE, H. L. K., AND P. J. HASTINGS. 1965. The analysis of genetic recombination on the polaron hybrid DNA model. Genet. Res. 6:27–92.

WILSON, G. B., A. H. SPARROW, AND V. POND. 1959. Sub-chromatid rearrangements in *Trillium erectum*. I. Origin and nature of configurations induced by ionizing radiation. Amer. J. Bot. 46:309–316.

WOLFF, S. 1965. On the chemistry of chromosome continuity. Nat. Cancer Inst. (U.S.A.) Monogr. 18, pp. 155–188.

SUPPLEMENTARY REFERENCES

JOHN, B., AND K. R. LEWIS. 1969. The chromosome cycle. Protoplasmatologia 6(B):1–125.

PEACOCK, W. J., AND J. R. D. BRECK (eds.). 1968. Replication and recombination of genetic material. Australian Academy of Science, Canberra.

TAYLOR, J. H. (ed.). 1967. Molecular genetics. Academic, New York.

———. 1969. The structure and duplication of chromosomes, pp. 163–221. *In* E. W. Caspari and A. W. Ravin (eds.), Genetic organization, vol. 1. Academic, New York.

VALENCIA, J. I., AND R. F. GRELL (eds.). 1965. Genes and chromosomes: structure and function. Nat. Cancer Inst. (U.S.A.) Monogr. 18.

WAGNER, R. F. (ed.). 1969. Nuclear physiology and differentiation. Genetics (Suppl.) 61:1–469.

2 MITOSIS, MEIOSIS, AND FERTILIZATION

The vegetative multiplication of cells is usually correlated with the multiplication of nuclei. The products of nuclear division, however, need not be distributed to different cells. Nuclear division is distinguishable from cell division, and when both processes are synchronized, nuclear and cellular division is termed *mitosis*. The events occurring during nuclear division captured the attention of the early cytologists, and the localization of the genes in the chromosomes so fascinated the early geneticists that mitosis has generally been described only in terms of nuclear and chromosomal observations.

The sexual reproduction of animals was related first to certain specialized cells, the gametes, and then to their origin. For plants, the alternation of generations in the lower groups had to be extended to the higher groups before the role of the sporophyte and gametophyte in sexual reproduction could be fully appreciated. The sporophyte produces the gametophytes, which, in turn, are responsible for the production of the gametic nuclei. Once the sporophyte was identified as the diploid generation and the gametophyte as the haploid generation, the basic similarity of the nuclear and chromosomal events in the sexual reproduction of animals and plants was revealed. In terms of nuclei and chromosomes, fertilization results in the pooling of chromosomes in the haploid nuclei in gametes or gametophytes to yield the zygotic nucleus.

The careful study of sectioned, stained germinal tissue of animals and plants during the last century was responsible for the discovery of nuclear and chromosomal phenomena leading from a diploid cell to a haploid gamete or gametophyte and for the determination of their sequence. The process was termed *meiosis* and was described almost completely in nuclear and chromosomal terms. A rather comprehensive picture of mitosis and meiosis had been drawn before the rediscovery of the Mendelian laws. Boveri in 1895 explained the different types of development of individuals resulting from the unorthodox distribution of chromosomes in treated, fertilized sea urchin eggs by assuming that the chromosomes might somehow be involved in heredity and development. Cell lines with different chromosome numbers were associated with a characteristically aberrant type of development.

Mitosis and meiosis are dynamic processes in which a number of nuclear, chromosomal, and cellular events mesh in time and space. Mitosis is essentially a means for making nuclear carbon copies: one nucleus yields two nuclei identical to each other and to their progenitor in both chromosome number and morphology. Assuming that a cell has a structurally altered chromosome or chromosome number and can undergo mitotic division, the aberration usually will be transmitted to the daughter nuclei. In meiosis, the aberration can lead to additional and perhaps different aberrations. In some cases, an aberration in meiotic cells will be responsible for malfunctioning or inviable gametophytes, zygotes, or embryos. For certain chromosomal aberrations, the genetic consequences can have a selective value, so that the aberrant chromosomes are incorporated into the chromosome complement as a permanent characteristic of a population or species. Fortunately, many types of chromosomal aberrations are transmitted from one generation to the next and provide the raw materials of cytogenetics.

MITOSIS

Although plant and animal cells display certain structural differences during mitosis, the sequence of nuclear and chromosomal (if not cytoplasmic) events is amazingly similar, and from a cytogenetic point of view, significant differences are not observed. The nucleus in the premitotic cells of plants and animals usually cannot be distinguished from a nucleus which will not undergo mitosis. While such names as *resting* or *metabolic nucleus* have shortcomings, usage seems to favor the former term.

The events during mitosis were first described for killed, fixed, sectioned, and stained tissues or cells by cytologists who were unable to detect changes in the cytoplasm but could readily observe significant nuclear and chromosomal events. Mitosis was arbitrarily divided into stages, each characterized by a number of easily identified landmarks. The division of mitosis into clearly defined stages should not be misconstrued as indicating pauses in a dynamic orderly process.

Resting Nucleus

A properly stained resting nucleus exhibits a chromatinic network, an artifact resulting from stained precipitated DNA and protein (Fig. 2-1). Except for a few species, the chromosomes cannot be detected. The resting nucleus generally offers little information to the cytogeneticist interested either in the number or the morphology of the chromosomes. In some species, several darkly staining, irregular bodies are scattered throughout the nuclear volume. These bodies, termed *chromocenters,* usually correspond to chromosomal segments which do not become tenuous and therefore stand out in the network. In most diploid species, the resting nucleus exhibits one or

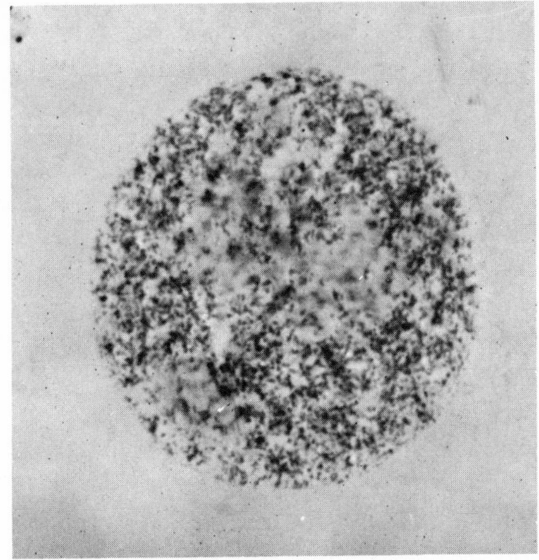

Fig. 2–1. A resting somatic nucleus in *Paeonia californica*. In this preparation, the nucleoli were not stained. (*Photograph courtesy of Dr. M. S. Walters and Dr. S. W. Brown.*)

two heavily staining bodies, the *nucleoli*. Although the nucleolus has long been a valuable chromosomal marker for certain cytogenetic studies, this nuclear organelle has recently received considerable attention from cell biologists interested in RNA synthesis.

Heitz (1931) first reported a correlation between the number of nucleoli in the resting nucleus and the number of certain chromosomes, the nucleolus-organizing chromosomes. A careful study of these chromosomes at the appropriate stage of mitosis or meiosis indicated that the nucleolus is organized at a specific site. In diploid species with one pair of nucleolus-organizing chromosomes, the resting nucleus has either two nucleoli or one nucleolus with a larger volume. McClintock (1934) confirmed these observations in maize and characterized the nucleolus-organizing site as a chromosomal segment which could be broken to yield two shorter segments, each capable of producing a nucleolus. The nucleolus-organizing region, therefore, is a complex structure usually restricted to one or more pairs of chromosomes.

The dipteran genus *Chironomus* provides unusual material for a study of the inheritance of the nucleolus-organizing region (Beerman, 1960). In *C. tentans*, two different chromosomes have a nucleolus-organizing region, so that up to four nucleoli are found in the resting nucleus; and in *C. pallidivittatus*, one of the chromosomes has a nucleolus-organizing region but at a different site (Fig. 2-2). By appropriate crosses among the progeny from the fertile interspecific hybrid, recombinants were obtained with zero to six nucleoli in the resting nuclei of embryos. Although recombinants with one to six nucleoli develop normally, recombinants without a nucleolus die at the embryonic stage.

One mutant in the toad *Xenopus laevis* was lethal for tadpoles when homozygous and furnished suitable material for investigating the role of the nucleolus in cell metabolism.

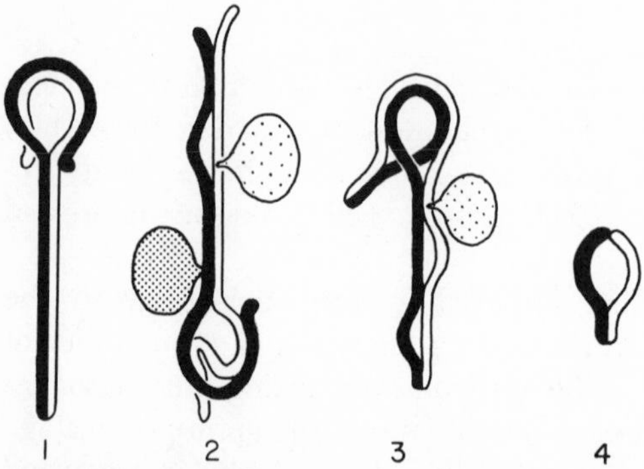

Fig. 2–2. Pairing of the salivary-gland chromosomes in the hybrid between *Chironomus pallidivittatus* (*black*) and *C. tentans* (*outline*). The one nucleolus of *C. pallidivittatus* and the two nucleoli of *C. tentans* are represented by hatched circles. (*John and Lewis, 1963; after Beerman, 1960.*)

The homozygous mutant tadpoles lack nucleoli because the organizing region is missing from the nucleolus chromosomes. By using pulses of RNA precursors and noting the presence or absence of labeled ribosomal and of transfer RNA in homogenates from the normal and mutant tadpoles, the nucleolus was implicated in the synthesis of 18S and 28S ribosomal RNA (Brown and Gurdon, 1964). Later experiments also indicated a complex nucleolus-organizing region in biochemical terms: this segment contains the DNA for ribosomal RNA, and each cistron coding for the ribosomal subunits may have up to 1,600 copies.

During nuclear division, the chromosomes become visible at prophase and, at the end of telophase, later merge into the chromatinic network. In some species, darkly staining irregular bodies are present and, by 1908, they were no longer considered to be artifacts and were termed *prochromosomes* or *chromocenters*. By 1910, their number was correlated with the number of chromosomes in the nucleus. The relatively long, darkly staining segments adjacent to the centromere of the chromosomes was assumed to be responsible for the prochromosomes. Heitz (1929) called attention to the differential staining of the chromosome; the darkly staining material was termed *heterochromatin* and the lighter staining material, *euchromatin*.

Heterochromatin is generally considered to be genetically inert, that is, devoid of Mendelian (*A*, *a*) genes. In comparing the sites of genes in linkage maps for chromosomes with heterochromatic and euchromatic segments in such species as *Drosophila melanogaster* and *Lycopersicon esculentum* (tomato), the loci almost always occur in the euchromatic segments. Yet, heterochromatin can have an effect on gene action, chromosome behavior, or the phenotype. For example, the terminal heterochromatic segment on the abnormal chromosome 10 in maize is responsible for the neocentric activity of the heterochromatic knobs of the chromosomes in this species. Goldschmidt (1955) has stated: "One thing may be said with certainty: heterochromatin must play a considerable role in

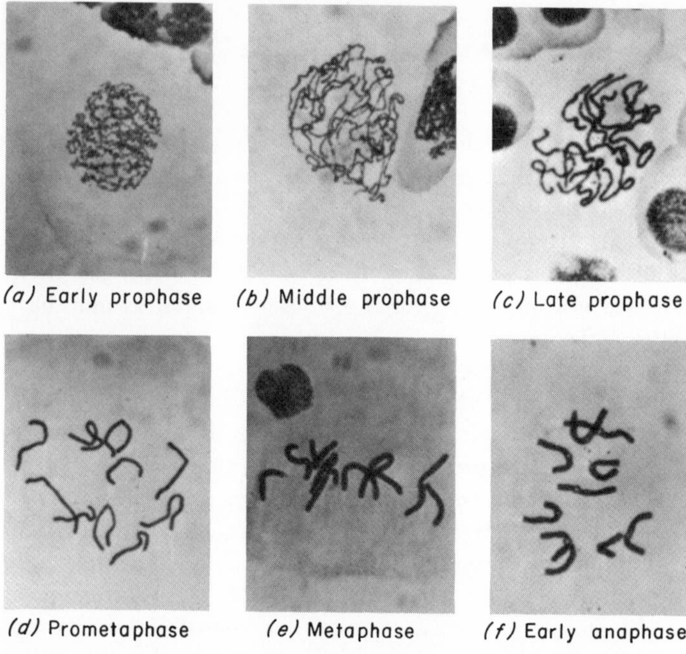

Fig. 2–3. Mitosis in *Paeonia californica*, $2n = 10$. The spindle apparatus and the nucleolus are not visible in the photographs. (*a*) Early prophase. (*b*) Middle prophase, showing two chromatids in the chromosomes. (*c*) Late prophase, with the clearly two-stranded chromosomes showing a poorly staining segment on the centromere. In this species, the 10 chromosomes can be counted at late prophase. (*d*) Prometaphase, showing two chromosomes with a long stalk and satellite and two chromosomes with a short stalk and satellite. Several chromosomes have a sharp bend at the centromere. (*e*) Metaphase. (*f*) Early anaphase.

the history of the chromosomes and in their function, and an important but unorthodox genetic role is expected."

In mammalian species, the resting nuclei of somatic cells from the female but not the male display a very small, darkly staining sphere, the Barr body, representing an inactivated X chromosome. Mammalian cytogeneticists use the number of these bodies as an indicator of the number of X chromosomes in aneuploids for these chromosomes.

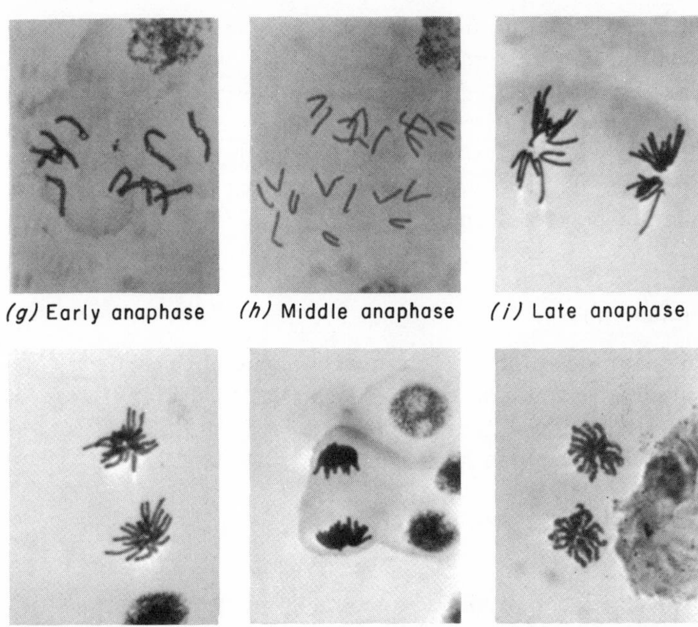

Fig. 2–3. (*Continued*). (*g*) Early anaphase. (*h*) Middle anaphase. (*i*) Late anaphase. (*j*) Early telophase. (*k*) Middle telophase. (*l*) Late telophase, merging into resting nucleus. (*Photographs courtesy of Dr. M. S. Walters and Dr. S. W. Brown.*)

Prophase

The onset of nuclear division is difficult to establish with certainty in chromosomal terms. The first indication of division is the gradual disappearance of the chromatinic network coincidental with the emergence of the long thready chromosomes (Fig. 2-3*a*). Throughout prophase, the chromosomes become progressively shorter and thicker as the result of coiling, as though each chromosome were a spring enclosed within a matrix (Fig. 2-3*b* and *c*). By midprophase, each chromosome can be seen to have two chromatids. In many species, each chromosome has a poorly staining segment which will be the site associated with fibers from the spindle apparatus at metaphase. Of all the terms proposed for this site, *centromere* has received widest acceptance in the United States while *kinetochore* is more frequently used in other countries. In carefully stained chromosomes, the chromatids can be seen to continue through the centromere. The nuclear membrane and the nucleoli are present throughout prophase.

Prometaphase

In plant cells, the nuclear membrane disappears before the spindle apparatus is detected. This relatively brief stage of mitosis has been termed *prometaphase* (Fig. 2-3*d*) and is usually not found in small samples of dividing cells.

Metaphase

The alignment of the chromosomes in a plane formed by their centromeres midway between the poles of the spindle apparatus characterizes metaphase (Fig. 2-3*e*). The spindle apparatus is essentially a complex of fibers extending from pole to pole and from poles to centromeres. At this stage, each centromere is associated with fibers from opposite poles. The chromosomes are generally scattered throughout the equatorial plane formed by the centromeres.

The early cytologists examined sectioned tissues for cells at metaphase to determine the number and morphology of the chromosomes. In many species with relatively long or numerous chromosomes, the serial sections cut through chromosomes or even both arms of one chromosome so that an accurate count was difficult. Once techniques were developed to separate and then to spread the cells in the single plane of a slide, metaphase chromosomes were readily counted and described in detail.

Anaphase

The separation of the sister chromatids of each chromosome at the centromere signals the beginning of anaphase (Fig. 2-3*f* and *g*). In species with relatively few chromosomes, the midanaphase chromosomes can be counted and described (Fig. 2-3*h*). The position of the centromere can often be determined by comparing the relative arm lengths for each chromosome. As the two groups of chromosomes, each with the same number of chromosomes, proceed to opposite poles, they are confined within the cone of the spindle, occupying progressively smaller volumes throughout anaphase (Fig. 2-3*i*).

Telophase

Before reaching their respective pole, and presumably at a critical volume, each group of chromosomes is enclosed within a nuclear membrane (Fig. 2-3*j*), and the spindle apparatus gradually disappears. The nucleoli seen at earliest telophase become increasingly larger. Within each nucleus, the chromosomes which started elongating in anaphase become progressively longer and thready (Fig. 2-3*k*), until the telophase nuclei merge into resting nuclei (Fig. 2-3*l*). In one sense, telophase seems to reverse the changes in chromosome morphology observed during prophase.

MEIOSIS

Sexual reproduction in eukaryotes involves the production of nuclei with the haploid chromosome number and the means for delivering these nuclei to effect a fertilization, thereby restoring the diploid chromosome number in the zygotic nucleus. The production of haploid nuclei from diploid cells involves a process, termed *meiosis,* which follows a basic pattern for most eukaryotes. The devices to accomplish fertilization in animals, however, differ markedly from those in plants. The immediate products of meiosis in animals are the gametes and in plants, the spores. For animals, spermatogenesis yields sperm, and oogenesis yields ova; for plants, microsporogenesis yields microspores, and megasporogenesis yields megaspores. In terms of sexual differences, sperm and microspores play a similar role, as do the ova and megaspores; that is, the sperm and microspores furnish the male gametic nucleus, and the ova and megaspores furnish the female gametic nucleus.

A basic and significant difference between higher plants and animals concerns the immediate products of meiosis and their relation to breeding data. The haploid microspores and megaspores develop into haploid individuals, the gametophytes, which in turn produce the gametic nuclei. When chromosomal aberrations or mutant genes impair the normal development of the gametophytes, they are inviable or malfunctional, so that fertilization cannot occur. Chromosomal aberrations or mutant genes generally do not interfere with the function of gametes, and fertilization can produce inviable or malfunctional zygotes.

Mitosis and meiosis differ with respect to the nuclear events and their products. In diploid species, mitosis yields two diploid nuclei, identical to each other and to their progenitor in genotype and in the number and morphology of the chromosomes; meiosis produces four haploid nuclei from one diploid nucleus. Mitosis is completed after one nuclear division and meiosis after two nuclear divisions. Finally, mitosis is usually associated with the vegetative multiplication of nuclei, cells, or organisms, while the products of meiosis are indirectly or directly involved with sexual reproduction.

Although the stages of meiosis have been assigned the same terms as their equivalents in mitosis, Roman numerals distinguish the stages in each nuclear division. Certain chromosomal events, found only during prophase I, have no counterpart during mitotic prophase. A generalized account of meiosis will be presented in terms of an idealized cell or meiocyte, and differences between plant and animal meiocytes will be noted at the appropriate stages.

Prophase I

The landmarks identifying each substage of prophase I are related to the morphology or behavior of the chromosomes within the nuclear membrane and have provided the terms to identify each substage: leptotene (thin thread), zygotene (yoked thread), pachytene (thick thread), diplotene (double thread), and diakinesis, essentially the end of both diplotene and prophase I. Certain events in the substages have cytogenetic significance and will be discussed in some detail.

Leptotene

The chromosomes emerge from the chromatinic network as very long, thready strands, obviously longer and threadier than the chromosomes at earliest mitotic prophase (Fig. 2-4*a*). The chromomeres are readily observed as small, irregularly shaped and spaced, heavily staining material. During this substage, the chromosomes of most animal and some plant species appear to be associated with the nuclear membrane in a remarkable manner. The ends of each chromosome are appressed to a restricted area of the membrane opposite to the centrioles to form a bouquet. Some models to explain the pairing of homologous chromosomes have assumed that the bouquet might be involved in the orientation of the homologous chromosomes. In most plant and some animal species, the chromosomes form an irregular mass, misleadingly termed the *synizetic knot*, which is apparently associated with the nucleolus.

Zygotene

The onset of this substage is marked by the lateral associations of segments of two chromosomes (Fig. 2-4*b*). This pairing,

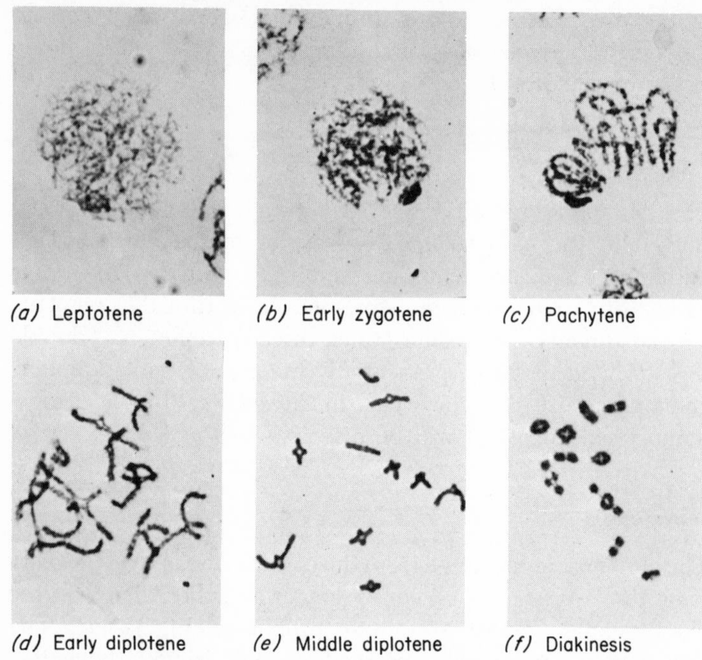

Fig. 2–4. Spermatogenesis in a male elaterid beetle (*Agriotes mancus* Say) with nine bivalents + XO. In this species, as is often the case in beetles, the prophase II stage is not well defined, and the cells quickly proceed from telophase I through interphase to metaphase II. (*a*) Leptotene, showing the precociously condensing X chromosome. (*b*) Early zygotene, a side view of the developing bouquet. (*c*) Pachytene, the bouquet of nine bivalents and the pear-shaped X chromosome. (*d*) Early diplotene. (*e*) Middle diplotene. (*f*) Diakinesis, showing the quadripartite nature of the acrocentric and metacentric chromosomes.

or *synapsis,* is the visible expression of the homology of the associated segments. A careful examination of the paired segments in favorable material reveals a precise pairing of the chromomeres in each segment.

The usual interpretation of the zygotene observations assumes that the single-stranded leptotene chromosomes first pair at homologous segments. According to Moens (1964),

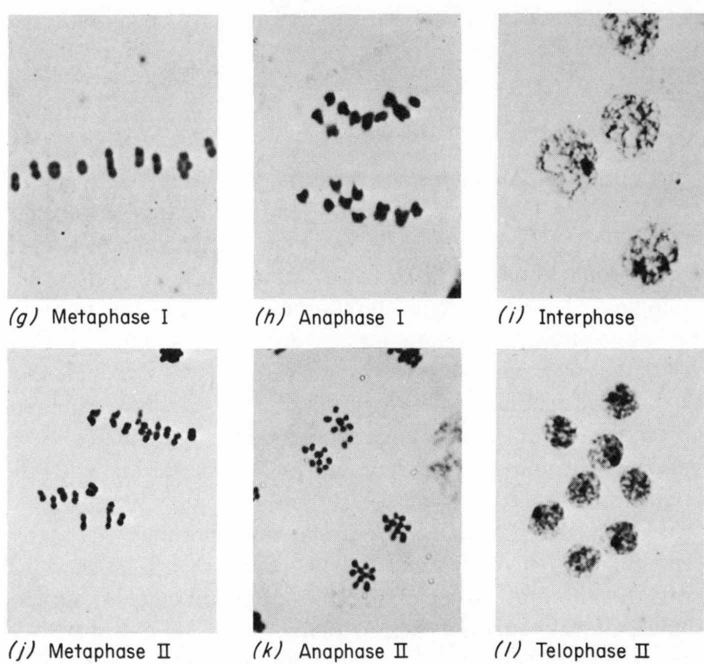

Fig. 2–4. (*Continued*). (*g*) Metaphase I, showing the X chromosome off the equatorial plane. (*h*) Anaphase I, with nine chromosomes in one group and ten chromosomes in the other group. (*i*) Interphase, one cell with the heterochromatic X chromosome. (*j*) Metaphase II. (*k*) Anaphase II. (*l*) Two sets of four cells at telophase II with two of the four cells in each set displaying the heterochromatic X chromosome. (*Photographs courtesy of Dr. S. G. Smith; Smith, 1956.*)

prophase I starts with leptotene and proceeds directly to pachytene. Other cytologists claim that pairing occurs in the premeiotic diploid nucleus and that synapsis at pachytene plus the eventual appearance of the chiasmata are devices to ensure the normal distribution of the homologous chromosomes during the first meiotic division.

The pairing of homologous chromosomal segments continues throughout zygotene in a manner resembling a zipper. The occasional apposition of nonhomologous segments for chromosomes with structural aberrations has been explained by this zipper effect.

Pachytene

Once all the homologous segments have paired, the synapsed chromosomes are shorter and thicker than in the earlier substages (Fig. 2-4*c*). The pairing of all the segments in each chromosome indicates that the chromosomes are indeed homologous. The pachytene chromosomes provide the most favorable material to characterize each chromosome by their relative length, arm ratios (long arm/short arm), and such topological features as heavily staining segments, prominent chromomeres or knobs, and constrictions. The nucleolus-organizing chromosome is identified by its association with the nucleolus.

Cytogeneticists usually examine chromosomes at pachytene for evidence of structural aberrations and to establish both the site and extent of the aberration in specific chromosomes. Certain species, such as maize (Fig. 2-5 *a* and *b*) and tomato (Fig. 2-6*a* and *b*), with superb pachytene chromosomes have been successfully exploited for detailed cytogenetic studies.

While the light microscope does not show a structure associated with the pachytene chromosomes, the electron microscope has revealed a highly organized structure of filaments, the *synaptinemal complex* (see frontispiece), between the paired chromosomes. The discovery of the complex in crayfish spermatocytes by Moses (1956) has been confirmed in other animal and plant species. The complex appears as a triplet of parallel dense bands in a single plane, curving and twisting along the axis between the chromosomes. The association of the synaptinemal complex with pachytene but not mitotic chromosomes suggests some functional role in synapsis, crossing-over, or chiasma formation. These speculations are supported by the observation that individual meiotic chromosomes (univalents) are not associated with a recognizable complex.

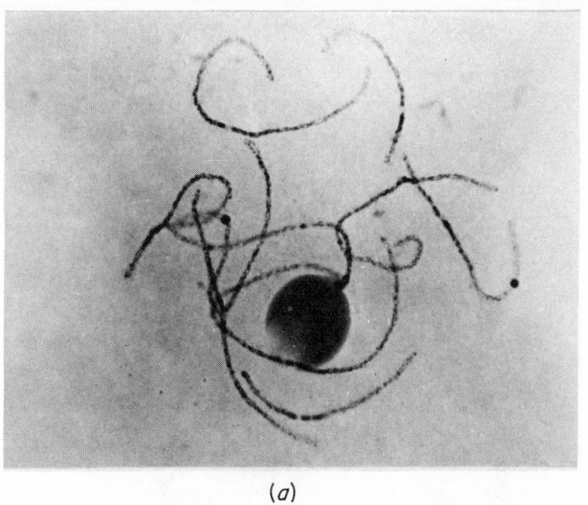

(a)

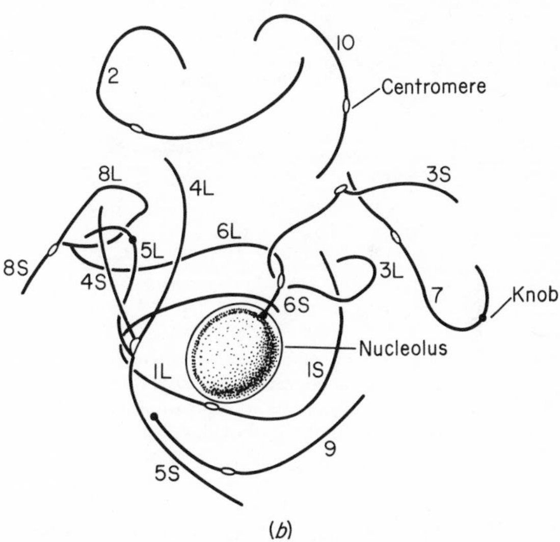

(b)

Fig. 2–5. (*a*) Pachytene chromosomes of maize, $n = 10$; (*b*) schematic diagram showing the site of the centromere in each chromosome. (*Photograph courtesy of Dr. D. T. Morgan, Jr.; Rhoades, 1950.*)

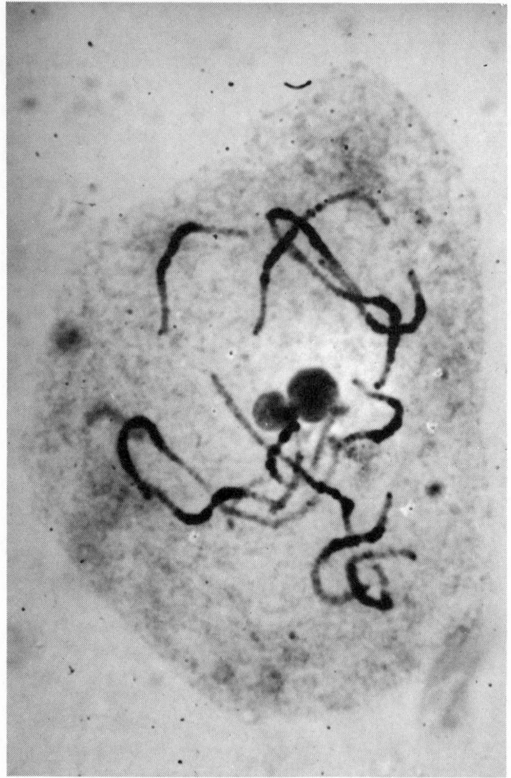

Fig. 2–6. Pachytene chromosomes of tomato ($n = 12$). Note the heterochromatic segments on either side of the centromere and the prominent terminal chromomere at the end of the chromosome arms. (*Photograph courtesy of Dr. D. W. Barton.*)

The oocytes but not the spermatocytes of *Drosophila melanogaster* exhibit synaptinemal complexes. In this species, crossing-over occurs in the female but not in the male. The recessive 3(C)G mutation is responsible for the absence of crossing-over, asynapsis, and a high frequency of nondisjunction in the mutant female. Smith and King (1968) found synaptinemal complexes in heterozygous but not homozygous mutant females. These observations indicate that the synapsis of homologous chromosomes may be correlated with the appearance of synaptinemal complexes, but it is not yet possible to assume that the complex is responsible for synapsis.

Diplotene

While the number of chromatids in the chromosomes prior to diplotene might not be clear, each diplotene chromosome has two chromatids and a single centromere. The synapsed homologous chromosomes separate from each other except where they are held together by the mutual switching of nonsister chromatids at one or more sites along the chromosomes (Fig. 2-4*d* and *f*). The site of a mutual switching of nonsister chromatids is termed a *chiasma.* The paired chromosomes, or bivalents, are shorter and thicker than at pachytene. When homologous chromosomes do not form at least one chiasma, they occur as univalents.

The obvious presence of two chromatids in each chromosome at diplotene was responsible for the early notion that chromatid replication occurred during pachytene. The genetic recombination of linked genes (crossing-over) is known to occur at the four-strand stage, that is, when each chromosome has two strands. Thus, it was reasonable to assume that crossing-over was the result of the breakage and fusion of nonsister chromatids at pachytene and that a chiasma was the visual consequence of crossing-over. The synthesis of DNA, however, has been shown to be essentially completed prior to diplotene. The synaptinemal complex might play a role in the meiotic crossing-over of synapsed homologous chromosomes.

Diakinesis

The chromosomes in the bivalents at earliest diplotene are shorter and thicker than at pachytene and become increasingly shorter and thicker throughout this substage. The changes in the length and diameter of the chromosomes are correlated with the progressively tighter coiling of each chromosome.

The end of diplotene (Fig. 2-4g) is termed diakinesis. Cytogeneticists usually determine chromosome associations at this substage when the bivalents or associations of more than two chromosomes are scattered throughout the nucleus and readily observed. The nucleolus-organizing chromosomes are still in contact with the nucleolus.

The number of chiasmata declines from earliest diplotene to diakinesis, a phenomenon known as *terminalization.* As each bivalent becomes increasingly shorter, the chiasmata move away from the centromere and eventually slip off their arms. Regardless of the number of chiasmata for each bivalent at earliest diplotene, at least one chiasma is present at diakinesis.

Metaphase I

In higher plant species, the loss of the nuclear membrane and the absence of the spindle apparatus identify the brief prometaphase I stage (Fig. 2-4*h*). The alignment of the bivalents midway between the poles of the spindle apparatus characterizes metaphase I (Fig. 2-4*i*). The centromere of each chromosome in the bivalents, however, is associated with fibers from only one pole—in contrast with the centromere of mitotic metaphase chromosomes, with fibers from opposite poles. Yet both the mitotic metaphase and metaphase I chromosomes have two chromatids.

Terminalization continues throughout metaphase I for chiasmata at sites between the centromere and the end of the chromosome arms (Fig. 2-7). Shorter bivalents usually lose their chiasmata before the longer chromosomes, and the homologous chromosomes of shorter chromosomes may proceed toward their respective pole while the longer chromosomes still appear as bivalents.

Anaphase I

In many species, the separation of the homologous chromosomes in the bivalents is usually synchronized. The loss of all

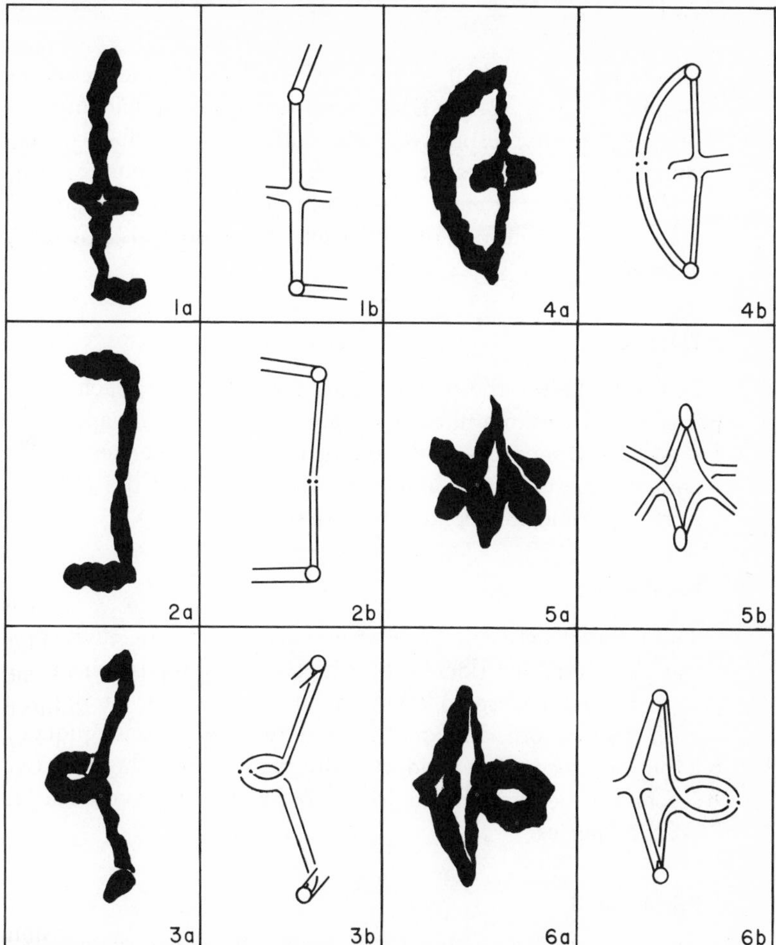

Fig. 2–7. Drawings of bivalents at metaphase I with diagrams to illustrate the number and site of chiasmata. (*1*) One nonterminal chiasma. (*2*) One terminal chiasma. (*3*) Two chiasmata in the same arm, one terminal and one nonterminal. (*4*) One terminal chiasma in one arm and one nonterminal chiasma in the other arm. (*5*) One nonterminal chiasma in each arm. (*6*) One nonterminal chiasma in one arm and one terminal and one nonterminal chiasma in the other arm. (*Lewis and John, 1964.*)

chiasmata for one or more bivalents marks the onset of anaphase I (Fig. 2-4*j*). Each group with the haploid number proceeds to its respective pole within the confines of the spindle and occupies an increasingly smaller volume (Fig. 2-4*k*). The arms of each chromatid separate but are associated with the one centromere, and each chromosome is obviously a *dyad.*

Telophase I

Sometime before the chromosomes reach the pole, each group is enclosed within a nuclear membrane. The spindle apparatus gradually disappears and usually cannot be detected at the end of this stage. The nucleolus reappears, and the chromosomes become longer and threadier.

Interkinesis

The interval between the first and second meiotic division is usually termed interkinesis (Fig. 2-4*l*) to distinguish this stage from the interphase following a mitotic division. At the end of the first meiotic division, the primary meiocyte has produced either one secondary meiocyte with two haploid nuclei or two secondary meiocytes each with a haploid nucleus, depending on the species.

Prophase II

The second meiotic division is essentially a mitotic division. Each nucleus usually enters prophase II at the same time (Fig. 2-4*m*) and synchronously proceeds through the subsequent stages. The chromosomal events observed during prophase I are not seen in prophase II.

Metaphase II

The nuclear membrane disappears, and the chromosomes are aligned in a plane midway between the poles of the spindle apparatus (Fig. 2-4*n*). While the centromeres of homologous chromosomes in bivalents at metaphase I are associated with fibers from only one pole, the centromeres of chromosomes at metaphase II are associated with fibers from opposite poles. Yet the metaphase I and II chromosomes have two chromatids. These observations suggest that the effective replication of the centromere may be out of phase with the replication of the chromatids during meiosis.

Anaphase II

The separation of the chromatids at the centromere signals the onset of anaphase II (Fig. 2-4*o*). Each chromatid has a centromere and therefore is a chromosome at this time. Each group with the haploid chromosome number proceeds toward its respective pole, and in species with relatively few, large chromosomes, the number and morphology of these chromosomes can be determined at midanaphase II.

Telophase II

Each group of chromosomes is enclosed within a nuclear membrane, and each spindle apparatus disappears. In plant species with a single secondary meiocyte, cell division in two planes yields four cells, each with a haploid nucleus. The chromosomes become longer and threadier; eventually the chromatinic network of the resting nucleus appears. If two nucleoli were present in the premeiotic nucleus, the haploid telophase II nuclei have a single nucleolus.

GAMETOGENESIS AND FERTILIZATION

The four haploid products of spermatogenesis in the male mature to yield sperm. During oogenesis, the first meiotic division occurs close to the cellular membrane, and one telophase I nucleus is extruded to form the first polar body. In the second meiotic division, the first polar body yields two second polar bodies, and the other haploid nucleus yields a third polar body and the egg nucleus. The course of nuclear events during oogenesis may be different in some species, but the end product is a haploid nucleus in the egg. In fertilization, the sperm delivers its haploid nucleus into the cytoplasm of the egg to give a zygote with the diploid chromosome number.

SPOROGENESIS AND FERTILIZATION

The four haploid nuclear products of microsporogenesis for the microsporocyte or pollen mother cell are packaged to give four viable uninucleate microspores. After the microspores separate, the microspore nucleus divides mitotically, producing a generative nucleus and a tube nucleus. In some species, these cells are shed as pollen grains and, after landing on the stigma, germinate to produce a pollen tube which grows in the direction of the female gametophytes in the ovary. The generative nucleus proceeds down the pollen tube and divides mitotically to produce the two haploid sperm nuclei (Fig. 2-8). In other species, the generative nucleus divides before

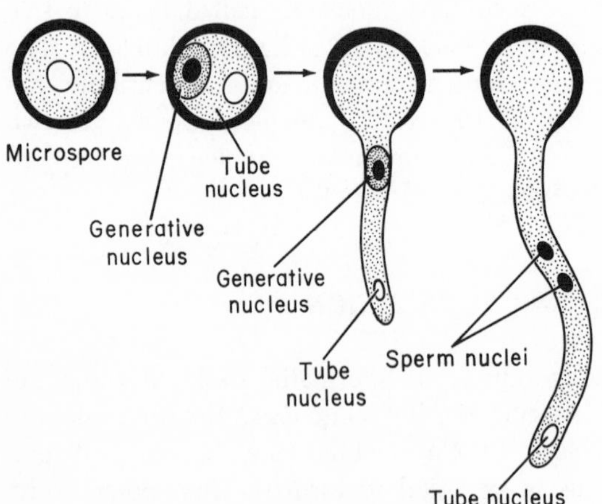

Fig. 2–8. Development of the male gametophyte (pollen grain) in higher plants.

the pollen grains are shed, and both sperm nuclei travel down the pollen tube. The trinucleate cell constitutes the mature male gametophyte in higher plants.

In most higher plants, each haploid product of megasporogenesis is enclosed within a megaspore. Three of the four megaspores forming a linear tetrad degenerate, and the basal megaspore develops into the female gametophyte, or embryo sac. In one common pattern, three successive mitotic divisions produce eight haploid nuclei so distributed that one becomes the egg nucleus and two become the polar nuclei (Fig. 2-9). Regardless of the pattern, the egg nucleus is haploid.

The pollen tube of the male gametophyte delivers the two sperm nuclei to the embryo sac, where one nucleus fertilizes the egg and produces the diploid zygotic nucleus. The other sperm nucleus fuses with the two polar nuclei, yielding a triploid primary endosperm nucleus, which divides mitotically in the development of the endosperm. The participation of the two sperm nuclei in the formation of a diploid zygotic

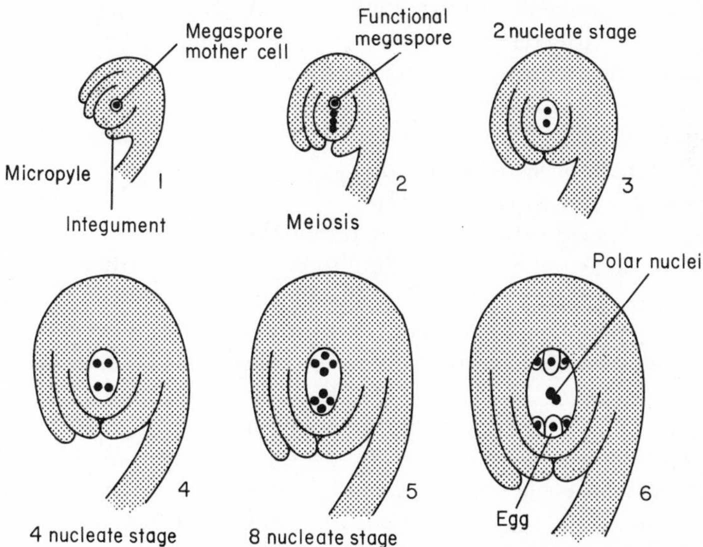

Fig. 2–9. The basic pattern of the deveolpment of the female gametophyte (embryo sac) in higher plants.

nucleus and a triploid primary endosperm nucleus has been termed *double fertilization.*

GENETIC CONTROL OF NUCLEAR EVENTS

The complex processes of nuclear division are usually integrated in both time and space to ensure the proper distribution of the chromosomes to daughter nuclei or cells. Mitosis and meiosis are not programmed processes which are automatically in phase once the nucleus or cell has been triggered. Moreover, the chromosomes are not merely passive vehicles for transporting genes and exempt from genetic effects on their morphology or behavior. Nuclear and cell division as well as chromosome morphology and behavior are determined by the genotype, and mutations can interfere with the normal mitotic and meiotic events.

Departures from the orderly course of meiosis can be expected to produce partial to complete sterility. Consequently, progenies segregating for sterility provide appropriate material to determine whether a genic mutation has altered either the sequence of meiotic events or chromosome morphology and behavior. Sterility can result from chromosomal aberrations, genic mutations, or environmental factors. When one-fourth of an F_2 population exhibits an obvious sterility, a recessive mutation is assumed to be responsible and meiocytes are examined for a meiotic aberration. Pollen abortion or ovule abortion as indicated by failure to set seed after pollination is readily detected in higher plants. Relatively few examples of genic mutations interfering with the orderly sequence of meiotic events have been found in animal species, and these were detected by aberrant genetic ratios. Most mutants have been found in *Drosophila*, as might have been anticipated from the volume of cytogenetic work in this genus. A selected number of genic mutations will be presented for plant species where cytological observations are available to determine the nature of the meiotic aberration (Table 2-1).

TABLE 2-1

Genetic Control of Meiotic Processes

Process	Species	Reference *
Pairing of homolog (asynapsis):		
No (or little) pairing	*Rumex acetosa* ♂	Yamamoto, 1934
	Hevea brasiliensis	Ramaer, 1935
	Hyoscyamus niger	Vaarama, 1950
	Alopecurus myosuorides	Johnsson, 1944
	Lycopersicon esculentum	Soost, 1951
Variable pachytene pairing	*Zea mays*	Beadle, 1930, 1933; Miller, 1963
	Triticum aestivum	Li, Pao and Li, 1945
	Picea albies	Anderson, 1947
Prevention of pairing between homoeolog		Sears and Okomata, 1958
	Triticum aestivum	Riley and Chapman, 1958
		Riley, 1960
Chiasma formation:		
Occurrence (desynapsis)	*Allium amplectans*	Levan, 1940
	Secale cereale	Prakken, 1943
	Ulmus glabra	Eklundh-Ehrenberg, 1949
	Triticum aestivum	Sears, 1954
		Kempanna and Riley, 1962
Frequency	*Allium fistulosum* × *cepa*	Maeda, 1942
	Agrostis hybrids	Jones, 1956
	Secale cereale	Rees, 1955, 1957
		Rees and Thompson, 1956

TABLE 2-1

Genetic Control of Meiotic Processes (*Continued*)

Process	Species	Reference *
Localization	*Allium fistulosum* × *cepa*	Levan, 1936, 1941
		Maeda, 1942
	Secale cereale	Emsweller and Jones, 1945
		Rees, 1955
Terminalization	*Secale cereale*	Rees, 1955
Chromosome coiling	*Matthiola incana*	Lesley and Frost, 1927
	Hordeum vulgare	Moh and Nilan, 1954
	Lolium perenne	Thomas, 1936
	Zea mays	Rhoades, 1956
Spindle formation	*Drosophila simulans*	Wald, 1936
	Lolium perenne × *Festuca pratensis*	Darlington and Thomas, 1937
	Zea mays	Clark, 1940
	Clarkia exilis	Vasek, 1962
Orientation and disjunction	*Zea mays*	Beadle, 1932
		Rhoades and Vilkomerson, 1947
		Rhoades, 1952
	Secale cereale	Thompson, 1956
	Drosophila melanogaster	Gowen, 1933
		Lewis and Gencarella, 1952
		Spieler, 1963
Chromosome movement	*Drosophila melanogaster*	Lindsley and Novitski, 1958
		Ptashne, 1960
Cleavage of cytoplasm	*Zea mays*	Beadle, 1932

* For references see the original.

Source: John and Lewis, 1965.

The pollen mother cells in the anthers have the diploid chromosome number, so that meiosis yields functional haploid microspores. In barley ($n = 7$), a mutant recessive gene is responsible for the formation of polyploid pollen mother cells with different chromosome numbers by fusion of two or more cells prior to meiosis (Smith, 1942). Numerous aborted microspores result, and anther development ceases shortly after the completion of meiosis. Mitotic divisions in root tips and in the somatic cells of the anthers, however, are normal.

During mitosis or meiosis, each chromosome has a characteristic length and morphology at each stage or substage of prophase I. In *Matthiola incana* (stock), the highly sterile Snowflake variety has obviously longer chromosomes at prophase I and metaphase I than the fertile varieties. A cytogenetic study of hybrids between the Snowflake and fertile varieties and of their segregating F_2 progenies indicated that the abnormal length of the chromosomes is determined by a recessive mutant gene (Lesley and Frost, 1927). In barley, a recessive mutation is responsible for both shorter than usual pachytene chromosomes and for the precocious terminalization of chiasmata (Moh and Nilan, 1954). The high frequency of univalents at diakinesis and metaphase I in mutant plants leads to aborted aneuploid spores. The mitotic chromosomes of the mutant, however, are indistinguishable from those in normal plants.

During meiosis, homologous chromosomes synapse, form bivalents, and proceed through the nuclear divisions without adhering to each other. In some species, nonhomologous pachytene chromosomes may be associated at the heavily staining segments adjacent to the centromere, but these chromosomes cleanly separate at the later substages. An increase in the ambient temperature during meiosis can cause stickiness of the chromosomes. A recessive mutation in maize, however, is responsible for sticky chromosomes during meiosis and mitosis (Beadle, 1932*b*). The sticky meiotic chromosomes are attenuated and break during the anaphase stages to yield deficient chromosomes and acentric fragments, which produce aborted spores and low fertility.

Occasional univalents are not uncommon at diplotene, diakinesis, or metaphase I and usually have no genetic basis. In a number of plant species, high levels of sterility are caused by a recessive gene responsible for varying numbers of univalents. A cytological study of such mutants usually reveals asynapsis, a failure of homologous chromosomes to pair at pachytene (Beadle, 1930), or desynapsis, the absence of chiasma formation after pachytene. The random distribution of univalents during the meiotic divisions yields numerous spores with more or less than the haploid chromosome number.

Except for oogenesis, when the spindle apparatus is normally eccentric, the apparatus usually forms in the middle of the cells, and the fibers converge at the poles. A recessive mutation has been found in maize for a divergent spindle apparatus so long that it must curve to fit the cell (Clark, 1940). As the anaphase I chromosomes approach the "poles," they diverge with the spindle apparatus and form multiple nuclei at telophase I. Each small nucleus is associated with its own divergent spindle apparatus at metaphase II, and from four to ten microspores with different chromosome numbers are produced at the end of the second meiotic division. Reciprocal crosses between mutant and normal plants indicate that the mutation is expressed during microsporogenesis but not megasporogenesis.

Meiosis is completed after two nuclear divisions. In *Datura stramonium* a recessive mutation suppresses the second meiotic division during sporogenesis for many sporocytes (Satina and Blakeslee, 1935). Consequently, crosses between a mutant seed parent and a normal pollen parent yield diploid and triploid progeny.

The formation of a cell plate after the first meiotic division in maize pollen mother cells produces two secondary meiocytes, and at the end of the second division, a second cell plate yields four microspores. A recessive mutation in this species suppresses cell-plate formation after the first meiotic division in 1 to 60 percent of the pollen mother cells and in most of the megasporocytes (Beadle, 1932*a*). If the first cell division does occur, a second division gives four microspores.

These and other examples of the genic control of meiotic and mitotic events in plant and animal species have provided interesting material for the cell biologist and electron microscopist. Cytogeneticists have been quick to exploit such mutants to obtain progeny with structural or numerical chromosomal aberrations without resorting to physical or chemical agents.

REFERENCES

BEADLE, G. W. 1930. Genetical and cytological studies of Mendelian asynapsis in *Zea mays*. Cornell Univ. Agr. Exp. Sta. Mem. 129.

———. 1932*a*. A gene in *Zea mays* for failure of cytokinesis during meiosis. Cytologia 3:142–155.

———. 1932*b*. A gene for sticky chromosomes in *Zea mays*. Z. Induktive Abstammungs- Vererbungslehre 63:195–217.

BEERMAN, W. 1960. Der Nukleolus als lebenswichtiger Bestand del Zellkernes. Chromosoma 11:263–296.

BROWN, D. D., AND J. B. GURDON. 1964. The absence of ribosomal RNA synthesis in the anucleolate mutant of *Xenopus laevis*. Proc. Nat. Acad. Sci. 51:139–146.

CLARK, F. J. 1940. Cytogenetic studies of divergent meiotic spindle formation in *Zea mays*. Amer. J. Bot. 27:547–559.

GOLDSCHMIDT, R. B. 1955. Theoretical genetics. University of California Press, Berkeley.

HEITZ, E. 1929. Heterochromatin, Chromocentren, Chromomeren. Ber. Deut. Bot. Ges. 47:274–284.

———. 1931. Die Ursache der gesetzmässigen Zahl, Lage,

Form, und Grösse pflanzlicher Nukleolen. Planta 12:775–844.

John, B., and K. R. Lewis. 1963. Chromosome marker. J. & A. Churchill, London.

——— and ———. 1965. The meiotic system. Protoplasmatologia 6 (F/1):1–335.

Lesley, M. M., and H. B. Frost. 1927. Mendelian inheritance of chromosome shape in *Matthiola.* Genetics 12:449–460.

Lewis, K. R., and B. John. 1964. The matter of Mendelian heredity. Little, Brown, Boston.

McClintock, B. 1934. The relation of a particular chromosomal element to the development of the nucleoli in *Zea mays.* Z. Zellforsch. Mikroskop. Anat. 19:191–237.

Moens, P. B. 1964. A new interpretation of meiotic prophase in *Lycopersicon esculentum* (tomato). Chromosoma 15:231–242.

Moh, C. C., and R. A. Nilan. 1954. "Short" chromosome, a mutant barley induced by atomic bomb irradiations. Cytologia 19:48–53.

Moses, M. J. 1956. Chromosomal structure in crayfish spermatocytes. J. Biophys. Biochem. Cytol. 2:215–218.

Rhoades, M. M. 1950. Meiosis in maize. J. Hered. 41:58–67.

Satina, S., and A. F. Blakeslee. 1935. Cytological effects of a gene in *Datura* which causes dyad formation in sporogenesis. Bot. Gaz. 96:521–532.

Smith, L. 1942. Cytogenetics of a factor for multiploid sporocytes in barley. Amer. J. Bot. 29:451–456.

Smith, P. A., and R. C. King. 1968. Genetic control of synaptinemal complexes in *Drosophila melanogaster.* Genetics 60:335–351.

Smith, S. G. 1956. Spermatogenesis in an elaterid beetle. J. Hered. 47:1–10.

SUPPLEMENTARY REFERENCES

John, B., and K. R. Lewis. 1968. The chromosome complement. Protoplasmatologia 6(A):1–206.

MOSES, M. J. 1968. Synaptinemal complex. Ann. Rev. Genet. 2:363–412.

REES, H. 1961. Genotypic control of chromosome form and behaviour. Bot. Rev. 27:288–318.

RHOADES, M. M. 1961. Meiosis, pp. 1–75. *In* J. Brachet and A. E. Mirsky (eds.), The cell, vol. 3. Academic, New York.

3 UNUSUAL CHROMOSOMES

Because plant and animal chromosomes share a number of common characteristics, it is possible to consider a standard chromosome: a linear organelle with "Mendelian" genes, one centromere, two nonhomologous arms, and an allotment of genes oriented in a specific sequence. Nonstandard or unusual chromosomes, which have been found in individuals, all members of a species, certain groups of species, or a genus, provide valuable tools for certain cytogenetic studies.

The most commonly encountered unusual chromosomes are concerned with sex determination in animal species and a few dioecious plant species. The other types of unusual chromosomes are derived from standard chromosomes and are peculiar per se or common to the members of a species or genus. A ring chromosome has one centromere but no arms. Telocentric chromosomes, or *telochromosomes,* have a terminal centromere and only one arm; *isochromosomes* have homologous arms. Chromosomes may have two centromeres, no specific site of centromeric activity, or neocentric sites of activity. In a few species or hybrids in *Nicotiana,* some cells exhibit an extremely long chromosome termed a *megachromosome.* The supernumerary, or B, chromosomes appear to lack alleles for genes on the standard chromosomes and need not be present in the nuclei.

SEX CHROMOSOMES

Sex determination in animals and some plant species is directly or indirectly related to specific chromosomes. Usually, one sex has a pair of homologous chromosomes and the other sex a pair of heteromorphic chromosomes, one of which is the same as in the opposite sex. These chromosomes were formerly known as allosomes and the other chromosomes as autosomes. While the latter term has been retained, allosomes are now termed sex chromosomes. In some species, particularly insects with multiple sex chromosomes, their directed distribution during meiosis ensures the correct number and types of sex chromosomes in the gametes.

Animal species have been assigned to one of two groups, depending on the morphology of the sex chromosomes in the male and female. In one group, the female has homomorphic sex chromosomes and the male has heteromorphic chromosomes; in the second group, the situation is reversed. In either group, the sex with homomorphic chromosomes is homogametic, the other sex being heterogametic. The female in most vertebrate and many insect species is homogametic with two X chromosomes, and the male is heterogametic with one X and one Y chromosome. In birds, for example, the male is homogametic, and at one time, the sex chromosomes in these organisms were identified as Z or W chromosomes.

An extensive literature on the cytogenetics of sex chromosomes has accumulated. The three examples selected for discussion have had an impact on the general interests of cytogeneticists: sex determination in *Drosophila melanogaster,* in dioecious plant species, and in mammals.

Sex Determination in *Drosophila melanogaster*

The female is homogametic (XX) and the male heterogametic (XY). Sex-linked inheritance in this species was explained by assuming that the pertinent genes were carried in the X chromosome, and this assumption was validated by experimental evidence. Although the Y chromosome seems to be devoid of alleles for genes in the X chromosome, at least one allele and several factors for sperm motility have been found. The discovery of fertile polyploids, individuals with three or four sets of chromosomes, furnished the experimental material for investigating the role of the X chromosomes in sex determination for this species.

Progeny from appropriate crosses involving parents with different chromosome numbers were examined to determine their sex and the number of sex chromosomes and autosomes. By identifying males, females, or intersexes and determining their chromosomal constitution, a ratio of the number of X chromosomes to the number of sets of autosomes (X/A) correlated with sexual status (Table 3-1). A normal female (2X/2A) has a ratio of 1.0 and a normal male (1X/2A), a ratio of 0.5. The superfemales have a ratio beyond 1.0, and the supermales, a ratio less than 0.5. These terms do not refer to superfemininity or supermasculinity, and *metafemale* and *metamale* are currently used. Finally, flies with ratios between 0.5 and 1.0 are intersexes.

Bridges (1932) accounted for the interaction of the sex chromosomes and autosomes in sex determination:

> From the cytological relations seen in the normal sexes, in the intersexes, and in the supersexes, it is plain that these forms are based upon a quantitative relation between qualitatively different agents—the chromosomes. However, the chromosomes presumably act only by virtue of the fact that each is a definite collection of genes which are themselves specifically and qualitatively different from one another.

TABLE 3-1

The Number of Sex Chromosomes and Sets of Autosomes for the Different Recognized Sex Types in *Drosophila melanogaster*

Type	Chromosomes			X/A Balance
	X	Y	A	
Metafemale	3		2	1.5
Triploid metafemale	4		3	1.3
Female	4		4	1.0
Female	3		3	1.0
Female	3	1	3	1.0
Female	3	2	3	1.0
Female	2		2	1.0
Female	2	1	2	1.0
Female	2	2	2	1.0
Female	1		1	1.0
Intersex	3		4	0.75
Intersex	2		3	0.67
Intersex	2	1	3	0.67
Male	1		2	0.50
Male	1	1	2	0.50
Male	1	2	2	0.50
Male	1	3	2	0.50
Male	2		4	0.50
Metamale	1		3	0.33

Source: Gowen, 1961.

In other words, sex determination in *Drosophila* results from a genic balance among the chromosomes which can be tilted by the X chromosomes.

The role of the X chromosomes in the sex determination of *Drosophila* is not applicable to all insect species. In the hymenopteran species *Apis mellifera* (honeybee) and *Habrobracon juglandis,* the different sexes are not classified as homogametic or heterogametic. The diploid queen produces progeny from fertilized or unfertilized eggs. In the honeybee, the haploid male drones originate from unfertilized eggs and the occasional diploid males and female workers from fertilized eggs. To account for such observations in *Apis* and in *Habrobracon,* a locus with multiple alleles is held responsible for sex determination; homozygosity or hemizygosity (one allele) determines maleness, and a heterozygosity determines femaleness.

Sex Determination in Dioecious Plants

The male plant in *Melandrium album* is heterogametic (XY) and the female plant homogametic (XX). The Y chromosome is readily distinguished from the X chromosome and determines maleness, even in the presence of three X chromosomes, while its absence is responsible for femaleness. The number of autosomes has no apparent effect on sex determination in this species.

In diploid species of *Rumex* and *Humulus,* the female is homogametic (XX) while the male plant has multiple sex chromosomes (XY_1Y_2). During metaphase I, the three sex chromosomes in the male form a heterotrivalent, so that the flanking Y chromosomes proceed to the same pole and the middle X chromosomes to the opposite pole. The directed orientation of the heterotrivalent at this stage is not the usual situation for a trivalent of homologous chromosomes. In *R. acetosa,* genic balance determines sex, and the autosomes play a role as in *D. melanogaster.*

Although the sex chromosomes of *Spinacia oleracea* cannot be distinguished, breeding data indicate that the male behaves as though heterogametic (XY) and the female homogametic (XX). Sex determination, however, is accomplished by a series of alleles (Y, X^m, X) at a locus in one of the chromosomes. The XX genotype determines femaleness; the XY or X^mY genotypes, maleness. The situation in this species is reminiscent of the hymenopteran species.

Sex Determination in Mammals

The discovery of sex-linked genes in mammalian species and their mode of inheritance suggested that the male was heterogametic (XY) and the female homogametic (XX). For some years, sex determination in mammals was thought to be similar to the situation in *Drosophila,* but no experimental evidence was provided to test this assumption. The cytological demonstration of heteromorphic sex chromosomes (XY) in the human male and of homomorphic sex chromosomes (XX) in the female was shortly followed by the discovery of females with only one X chromosome and of males with one to four X chromosomes and one Y chromosome. These observations indicate that in this and other mammalian species, the Y chromosome is male-determining and its absence is female-determining, as in the dioecious plant species *Melandrium album.* A discussion of other characteristics of the mammalian sex chromosomes will be deferred until the chapter on mammalian cytogenetics.

No acceptable explanation has been offered for the origin of sex chromosomes in animal or plant species. It is reasonable to assume that the evolution of animals is somehow intimately related to the production of well-defined sexes. Mutations might have been responsible for diverting developmental processes so that one or the other sex would result. Consequently, a reasonably precise method for distributing maleness or femaleness genes would have been responsible for the distribution of these genes among the chromosomes and for the assumption by one chromosome of a "dominant" role. Another solution to the problem could lead to a single chromosome's becoming responsible for sex determination, so that its presence causes maleness and its absence femaleness.

Plant species presumably utilized developmental processes to determine the morphological sites of organs producing the different types of spores. Consequently, the separation of individuals into males or females may not have been a significant factor in the evolution of plants. For the comparatively few species which do have male or female individuals, sex chromosomes were necessary, and usually such chromosomes or their equivalents have been found. It is interesting to note that even these few species did not use a common method for sex determination.

RING CHROMOSOMES

Standard chromosomes do not form a ring. The ends of chromosomes are assumed to have a hypothetical structure, the *telomere*. When a chromosome has two breaks, yielding one centric segment with raw ends and two acentric fragments each with a telomere and a raw end, the fusion of centric or acentric segments is determined by a simple rule: only the raw ends can fuse. Ring chromosomes usually result from the fusion of the raw ends of the centric segment. When the ring is produced in this manner, the acentric fragments are eventually lost. Consequently, a cell or organism with a ring and a standard homologous chromosome is deficient for the genetic information in the acentric fragments. When a gametophyte receives the ring rather than the standard chromosome, the deficiency can lead to inviability; when a zygote has both ring and standard chromosomes, the deficiency may also produce inviability or malfunction. These possibilities are responsible for the comparative rarity of ring chromosomes, and, whenever found, the transmission of the ring is rather irregular and often restricted to the female gametophyte. Ring chromosomes have been observed in *Drosophila,* man, and several plant species and have been extensively studied in maize.

McClintock (1938) irradiated pollen grains of maize plants homozygous for the dominant allele of the brown midrib (*bm-1*) mutation on chromosome 5 with x-rays prior to pollinating mutant plants; 2 variegated plants were found among the 466 progeny. The brown stripes in these plants suggested that the dominant allele was occasionally lost in the heterozygous tissue. A cytological examination of the variegated plants revealed that each plant had a ring chromosome and a rod chromosome. One plant had a small ring (R-1) which compensated for the deficiency in a rod (Df-1) chromosome, and the other plant had a larger ring (R-2) compensating for the longer deficiency in a rod (Df-2) chromosome (Fig. 3-1). Furthermore, each ring and rod chromosome had a functional centromere. According to these observations, the centromere is a complex structure, which can be broken to yield functional fractional centromeres.

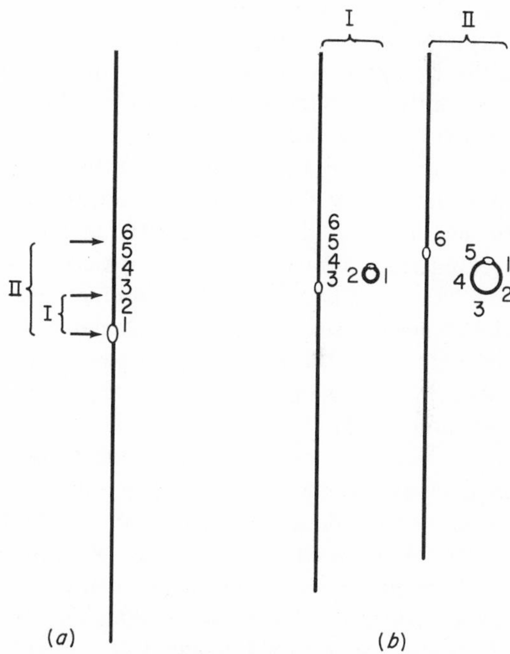

Fig. 3–1. (*a*) Diagram of a standard chromosome 5 in maize showing the sites of breaks (*bracketed arrows*) yielding a deficient rod chromosome and a compensating ring chromosome. The clear circle represents the centromere. (*b*) The deficient rod and ring chromosomes of I are represented as Df-1 and R-1, respectively, and those of II as Df-2 and R-2, respectively, in text. (*McClintock, 1938.*)

The mutant stripes in the heterozygous plants were explained by assuming that the dominant allele in the ring chromosome was lost during mitosis, thereby permitting the recessive phenotype to be expressed in the hemizygous cell (*pseudodominance*). An examination of pollen mother cells of anthers from different branches of the tassel indicated that some cells did not have a ring chromosome. Such cells presumably were derived from premeiotic cells lacking the ring chromosome. A cytological study of dividing root-tip cells showed that a cell could have the original ring chromosome, one larger or smaller ring chromosome, two interlocking ring chromosomes, two ring chromosomes, or no ring chromosome. Cells from the plant with the small R-1 chromosome more frequently exhibited two rings or no ring chromosomes than cells from the plant with the larger R-2 chromosome. On the other hand, cells from the plant with the R-2 chromosome more frequently displayed a dicentric ring which was larger or smaller than the original R-2 chromosome or two interlocking ring chromosomes at anaphase.

The dicentric large ring chromosome, two interlocking rings, and the small ring chromosomes in the root-tip cells of the plant with the large R-2 chromosome were explained by assuming a crossing-over between the sister strands of the R-2 chromosome (Fig. 3-2). While somatic crossing-over previously had been proposed to account for twin spots in *D. melanogaster* heterozygous for two linked genes (Stern, 1936), no genetic evidence compelled an explanation requiring an exchange between sister strands. Extensive linkage studies in *D. melanogaster* did not indicate any source of interference other than that attributable to nonsister-strand crossing-over. If a sister-strand crossing-over were responsible for an interference equivalent to that for nonsister-strand crossing-over, the recombination data for three linked genes might have indicated sister-strand crossing-over.

The extent of the variegation in the plants with a ring chromosome was determined by the size of the ring and by the locus of the mutant allele relative to the site of the centro-

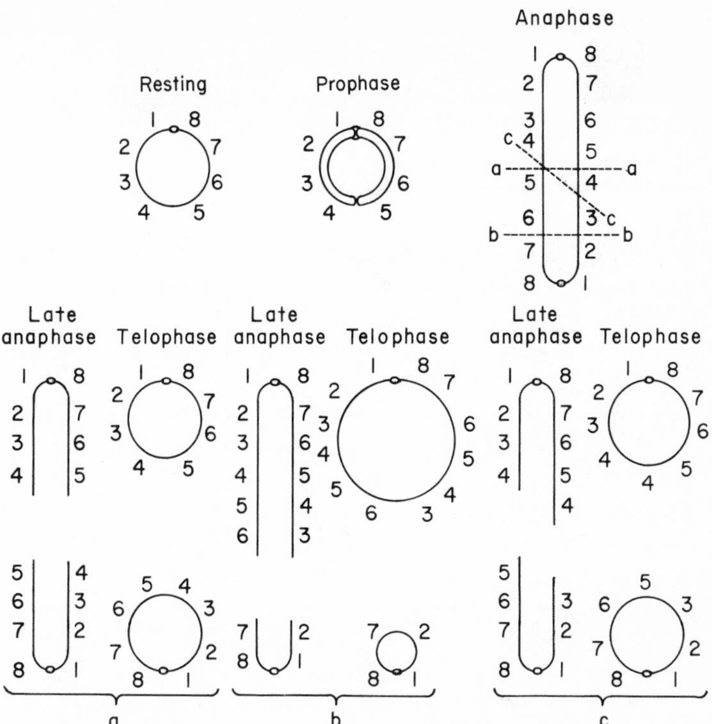

Fig. 3–2. *Upper left:* A ring chromosome with a centromere (*clear circle*) in a resting nucleus. The numbers represent the individual parts of the chromosome. *Upper middle:* Prophase configuration following crossing-over between sister chromatids of the ring chromosome. *Upper right:* The dicentric ring chromosome at the following anaphase. The intercentromeric strands can break at any position, three possible sites (*a, b,* or *c*) being indicated by broken lines. The broken strands at late anaphase and the new ring chromosomes formed by the fusion of the raw ends of these strands at telophase are presented in the bracketed figures for breaks *a, b,* and *c,* respectively. (*After McClintock, 1941.*)

mere in the ring. For the small R-1 ring, the degree of variegation was related to the frequency of cells which lost the ring during mitosis. For the larger R-2 ring, the frequency of brown stripes was related to the frequency of cells in which the

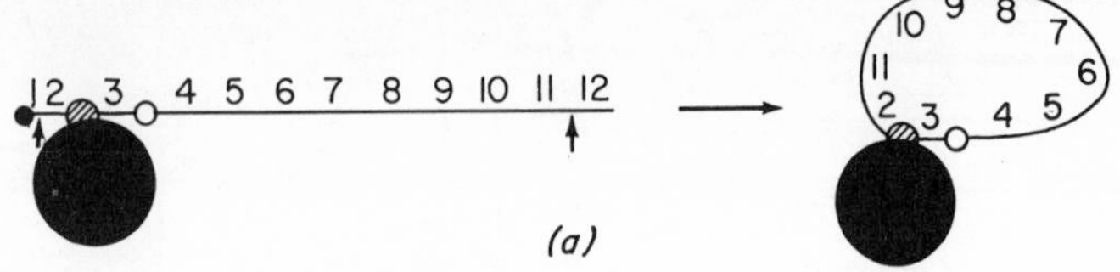

(a)

Crossover type | Anaphase I | Anaphase II

Single

Two-strand double

Three-strand double type I

Three-strand double type II

Four-strand double

(b)

double-sized dicentric ring chromosome was broken to yield two rings of different size. The centromeres of the dicentric ring chromosome proceeded to opposite poles, and when the ring was broken or cut by the cell plate, two rod chromosomes with raw ends were produced (Fig. 3-3). Depending on the sites of breakage, one cell may get a chromosome deficient for the dominant allele (*Bm-1*) and express the recessive phenotype. The rod chromosomes with broken ends appeared as ring chromosomes of different sizes in the next mitotic division.

The spores from a plant with a standard chromosome 5, a deficient rod chromosome 5, and the compensating ring chromosome may include (1) a standard chromosome, (2) a deficient rod and compensating ring chromosome, (3) a deficient rod chromosome, or (4) a standard and a ring chromosome. By using reciprocal crosses with a homozygous recessive standard plant, the functional gametophytes from the variegated plants could be detected by examining the variegated progeny. The deficient chromosome (Df-1 or Df-2) and the corresponding ring (R-1 or R-2) were present in the critical progeny. Furthermore, the Df-1 chromosome with the small deficiency but not the Df-2 chromosome with the larger deficiency was also detected in the recessive progeny.

Appropriate crosses between striped plants with Df-1 R-1 or Df-2 R-2 gave variegated progeny with Df-1 Df-2 R-2. The R-2 chromosome with the dominant allele for brown midrib-1 compensated for the deficient segment common to

---

Fig. 3–3. (*a*) Diagrammatic representation of the origin of the large ring chromosome following the breaks in the distal regions of chromosome 6 in maize. The nucleolus is represented by the large black circle, the centromere by the clear circle, and the nucleolus-organizing region by the hatched circle. The arrow indicates the probable sites of break induced by x-rays. (*Schwartz, 1953a.*) (*b*) Anaphase I and II configurations resulting from different types of crossovers between the standard and the ring chromosome. The arrows indicate the site of breakage in the bridge at anaphase I or II, and the terminal wiggle indicates the raw chromosome end at anaphase II. (*Schwartz, 1953b.*)

both the Df-1 and Df-2 chromosomes. When the R-2 chromosome was involved in a sister-strand crossing-over which led to a dicentric ring chromosome, the appropriate breaks in this chromosome gave smaller rings, so that the cell with such a ring chromosome had a deficiency for the dominant allele and expressed the recessive phenotype. Plants with Df-2 Df-2 R-2 R-2 had 22 and not the usual 20 chromosomes for maize and also were variegated for necrotic tissue. When cells lacked both ring chromosomes, they had a homozygous deficiency for the longer segment and degenerated before the surrounding cells matured. Consequently, ring chromosomes are one means of producing homozygous deficiencies for selected chromosomal segments in somatic cells.

Schwartz (1953*a*) investigated the cytological consequences of crossing-over in maize plants heterozygous for a large ring chromosome and the homologous standard chromosome and presented evidence indicating that sister-strand crossing-over can occur in meiotic chromosomes. Pollen grains with the dominant allele were irradiated with x-rays and used to pollinate plants homozygous for the recessive allele (*l-10, luteus*). One variegated plant appeared among the progeny and had a large ring and standard chromosome 6 (Fig. 3-3*a*). The viability of gametophytes with the large ring indicated that the deficient terminal segments were relatively small. The different chromosomal configurations at anaphase I and anaphase II in the pollen mother cells of the variegated plant are presented in Fig. 3-3*b*. By considering only single or double crossing-over in the long arm of the standard chromosome, it is possible to relate certain configurations at either anaphase to a particular type of double crossing-over. For example, the frequency of the type II three-strand double crossing-over can be determined from the frequency of cells at anaphase II with a single bridge, and cells with double bridges result only from a four-strand double crossing-over. According to the data (Table 3-2), 35 percent of the cell pairs at anaphase II had a single bridge, but only 13 percent of the cells at anaphase I had double bridges. These frequencies should have been approximately equal in the absence of chromatid interference, that is, assuming 1 two-strand : 2 three-strand : 1 four-strand double crossing-over. Moreover, negative chromatid interference did not explain the significantly higher than expected frequency of three-strand double crossing-over. Finally, 10 percent of the cell pairs at anaphase II had a double bridge, an observation which cannot be explained by single or double crossing-over. Sister-strand crossing-over was proposed to account for the high frequency of cell pairs with a single bridge at anaphase II and for cell pairs with a double bridge at this stage.

TABLE 3-2

Configurations at Anaphase I and II in Maize Plants Heterozygous for a Ring and Standard Chromosome 6 *

	Anaphase I				Anaphase II (Daughter-cell Pairs)			
	Single Bridge	Double Bridge	No Bridge	Total	Single Bridge	Double Bridge	No Bridge	Total
Number	368	81	171	620	166	47	262	475
Percent	59	13	28	100	35	10	55	100

* Note the significantly higher frequency of single bridges at anaphase II than of double bridges at anaphase I.

Source: Schwartz, 1953*b*.

A double crossing-over involving nonsister strands yields four types of configurations at anaphase: (1) a single bridge at anaphase I only from type I, three strands; (2) a single bridge at anaphase I and II from type II, three strands; (3) a double bridge at anaphase I from four strands; and (4) no bridge at anaphase I or II from two strands. A single bridge at anaphase II, however, can result from a crossing-over between nonsister chromatids in conjunction with a sister-strand crossing-over in the ring chromosome (Fig. 3-4). The ana-

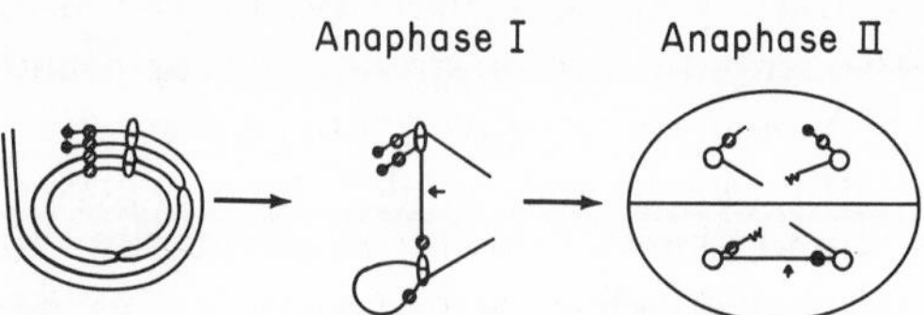

Fig. 3–4. Anaphase I and II configurations resulting from a single crossing-over between nonsister chromatids associated with one between the sister chromatids of the ring chromosome. The small arrows indicate a site of breakage in the dicentric chromatid bridge. (*Schwartz, 1953a.*)

phase II bridge from such a double crossing-over will account for the high frequency of the cells with an anaphase II bridge compared to the frequency of cells with double bridges at anaphase I. Furthermore, the unexpected double bridges at anaphase II can result from sister-strand crossing-over in the ring chromosome and either no nonsister chromatid crossing-over or two-strand nonsister double crossing-over.

The self-pollinated variegated plant with the ring chromosome gave 20 green, 25 variegated, and 71 luteus progeny. To account for the green progeny, Schwartz (1953*b*) proposed an ingenious manipulation of certain types of crossing-over. A single crossing-over produces a dicentric bridge at anaphase I, which breaks to yield three rod and one ring chromosome at anaphase II, but two rods have a raw end (Fig. 3-5*a*). When the bridge breaks at a site between the locus of the dominant allele and a centromere, one rod chromosome will have both the dominant and recessive alleles (duplication). Once this rod chromosome is included in the zygotic nucleus, the raw end heals and behaves like a telomere. Green plants with this duplication chromosome, however, can yield variegated plants with a ring chromosome, resulting from a four-strand double crossing-over involving the duplicated segment (Fig. 3-5*b*). A two-strand or a three-strand double crossing-over involving one strand of the ring chromosome will transfer the dominant allele from the ring to the standard homologous

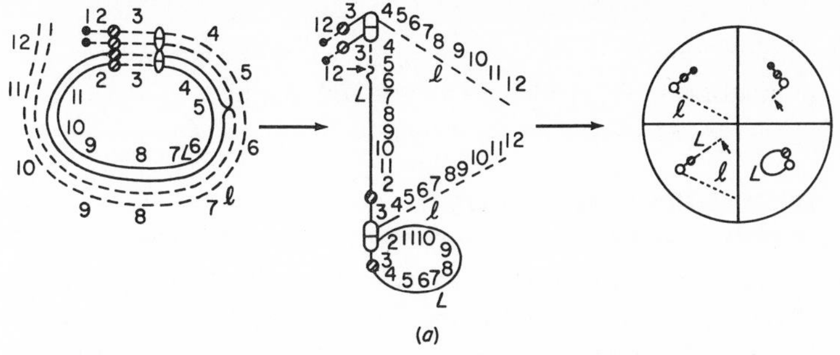

(a)

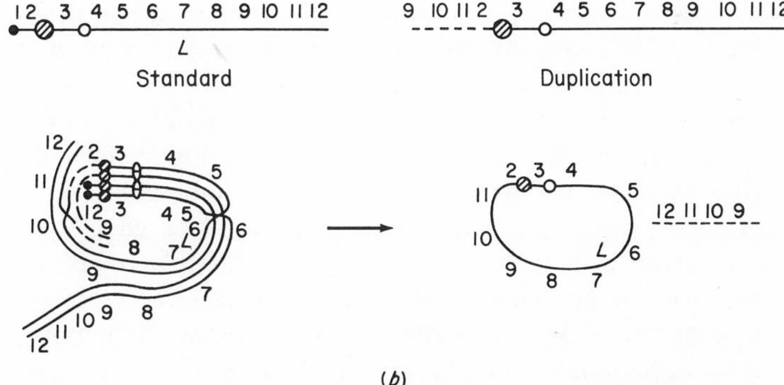

(b)

Fig. 3–5. The ring-rod-ring cycle in maize. (*a*) Diagram illustrating the origin of a rod chromosome with a duplicated segment and both the dominant (*L*) and recessive (*l*) alleles for the luteus mutation by a crossing-over between the homologous ring (*L*) and rod (*l*) chromosomes. (*b*) Diagram illustrating the origin of a ring chromosome from the duplication chromosome by four-strand double crossing-over between a standard and the duplication chromosome. The duplicated segment is indicated by the broken line. The meiotic pairing and the sites and chromatids involved in crossing over are shown in the bottom left diagram. The recovered ring with the dominant allele and the acentric fragment with the entire duplicated segment are shown in the bottom right diagram. (*Schwartz, 1953a.*)

chromosome to yield a green plant whose progeny after self-pollination will not include variegated plants.

TELOCENTRIC CHROMOSOMES

A standard chromosome has two nonhomologous arms, but one arm can be so short that the centromere appears to be terminal. Chromosomes with a truly terminal centromere are uncommon and presumably result from the misdivision or breakage of a standard chromosome at the centromere. In either case, the raw end heals when the chromosome is included in a zygotic nucleus. A telocentric chromosome, therefore, is deficient for one arm, and in diploid species, the deficient gametophyte or zygote is not likely to be viable or functional. The transmission of such chromosomes is determined by the frequency of their loss during meiosis, their conversion to another type of unusual chromosome (*isochromosome*) with two homologous arms, or their effect on the viability of gametophytes or zygotic nuclei. Furthermore, when a telocentric chromosome is an extra chromosome (*telocentric trisomy*) in a diploid plant species, the extra chromosome is usually responsible for the inviability or malfunction of some gametophytes or embryos.

Rhoades (1936) found a telocentric chromosome in the progeny of a maize plant with an extra chromosome 5 by noting that its phenotype was similar to that of the parental trisomic plant but not as extreme. The pachytene configuration indicated that the extra chromosome was indeed telocentric and derived from one arm of chromosome 5 (Fig. 3-6*a*). The telocentric chromosome was included in a heterotrivalent at metaphase I involving two homologous standard chromosomes (Fig. 3-6*b*). Approximately 50 percent of the pollen mother cells had the heterotrivalent, which was oriented at metaphase I so that the standard chromosomes proceeded to opposite poles at anaphase I.

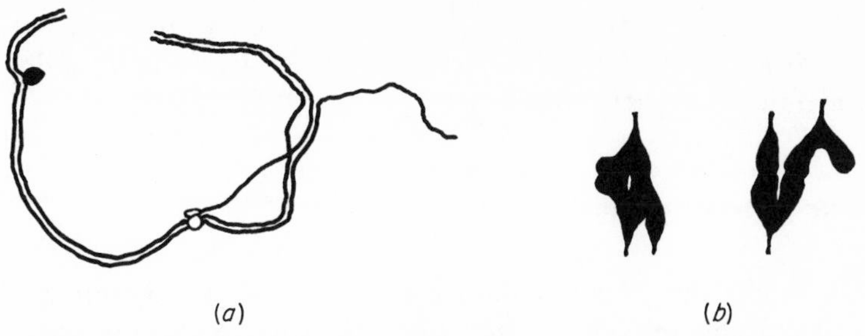

Fig. 3–6. (*a*) Camera lucida drawing of the synaptic relationships of the telocentric and two standard chromosomes 5 of maize at pachytene. The clear circles represent the centromeres; one standard chromosome has a deeply staining knob in the long arm. (*b*) Two heterotrivalents at metaphase I showing the orientation of the telocentric chromosome which permits the disjunction of the telocentric and one standard chromosome to the same pole. (*Rhoades, 1936.*)

The telocentric chromosome was present in 2 percent of the functional pollen grains and in 33 percent of the functional female gametophytes. The male gametophyte generally tolerates less imbalance than the female gametophyte with respect to chromosomal duplication or deficiency. Assuming that the frequency of the heterotrivalent is the same in the male and female meiocytes and a genotype of *Aa* telo-*A*, the frequency of megaspores with different genotypes can be calculated: 25 percent *A* and 25 percent *a* from meiocytes lacking the heterotrivalent and 12.5 percent *A* telo-*A*, 12.5 percent *a* telo-*A*, 12.5 percent *A*, and 12.5 percent *a* from meiocytes with the heterotrivalent. The 2,292 progeny from the testcross *Aa* telo-*A* × *aa* included 33.6 percent recessive progeny compared with the calculated value of 37.5 percent. The progeny from the control testcross *Aa* × *aa* included 50 percent recessive individuals.

Telocentric chromosomes have been useful tools in assigning a gene to one or the other arm of a standard chromosome and in mapping the centromere with precision, particularly when closely linked genes are in different arms. Khush and

Rick (1968) have used telocentric trisomics to great advantage in the tomato, the chromosomes of which are readily identified as pachytene.

ISOCHROMOSOMES

A misdivision of a standard chromosome at the centromere, an unstable telocentric chromosome, or a reciprocal translocation between homologous standard chromosomes at their centromere can produce an isochromosome with homologous arms. When the isochromosome is an extra chromosome, the numerical aberration is termed *secondary trisomy*, and the extra chromosome is called a *secondary trisome*. In a secondary trisomic, one chromosome arm is quadruplicated. The replacement of a standard chromosome by one of the corresponding isochromosomes leads to a deficiency for an entire arm which is usually lethal in diploid species.

Sen (1952) found two isochromosomes which could replace the corresponding standard chromosome in tomato. One isochromosome was identified as 8L·8L and the other as 9L·9L, indicating the long arm of chromosomes 8 and 9, respectively. Although the 9L isochromosome was poorly transmitted through the female gametophyte when it replaced the standard chromosome, transmission was increased when the isochromosome was a secondary trisome. The chromosome associations at diakinesis for the primary trisome 9L·9S and for the secondary trisome 9L·9L are summarized in Table 3-3. A ring of three chromosomes or a univalent with a chiasma in the homologous arms characterizes the secondary trisome.

Homologous acrocentric chromosomes with one long and one very short arm can be involved in a reciprocal translocation (a mutual exchange of segments but not a crossing-over) so that one translocation chromosome has the long arm and almost all of the other long arm. The other translocation chromosome with two short arms may be lost without any detectable effect on viability or function. The long translocation chromosome behaves like a pseudoisochromosome. The best known example of such a chromosome occurred in *D. melanogaster* (Morgan, 1922) and has been commonly termed an attached X chromosome. This chromosome played a significant role in understanding the genetics of double crossing-over.

TABLE 3-3

Frequency of the Different Types of Configurations at Diakinesis in a Secondary and a Primary Trisomic of Tomato

	Configuration								
Trisomic Type									**Total**
Secondary 9L·9L	3	16	16	9	2	1	2	1	50
Primary 9L·9S	19	. .	18	. .	. .	. .	9	4	50

Source: Sen, 1952.

Not all the four products from one meiocyte in higher plants and animals are recovered together, and the genetic consequences of crossing-over are deduced by a statistical analysis of the breeding data. In the females of *D. melanogaster* with an attached X chromosome, the two arms are involved in crossing-over, and two of the four chromatids are recovered in female progeny (Fig. 3-7*a*). When one arm has the dominant and the other arm the recessive alleles, crossing-over yields recessive progeny (*homozygosis*). The frequency of mutant progeny is a measure of crossing-over between each locus and the centromere. The closer a locus is to the centromere, the fewer the recessive progeny recovered. When a crossing-over between linked genes is studied, it is possible to obtain evidence pertinent to the question of sister-strand crossing-over (Beadle and Emerson, 1935). When sister-strand and reciprocal or nonreciprocal nonsister-strand crossing-over

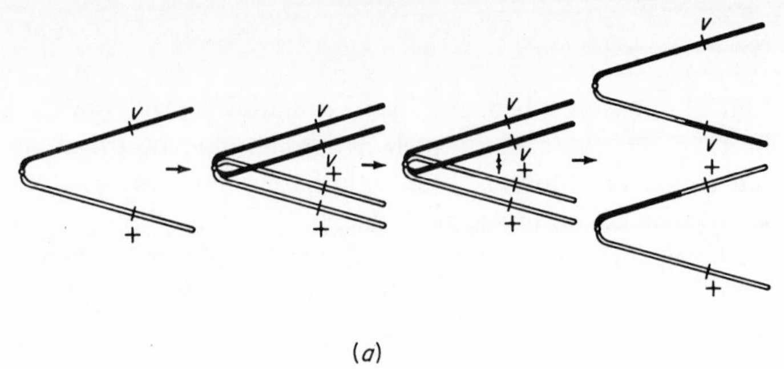

(a)

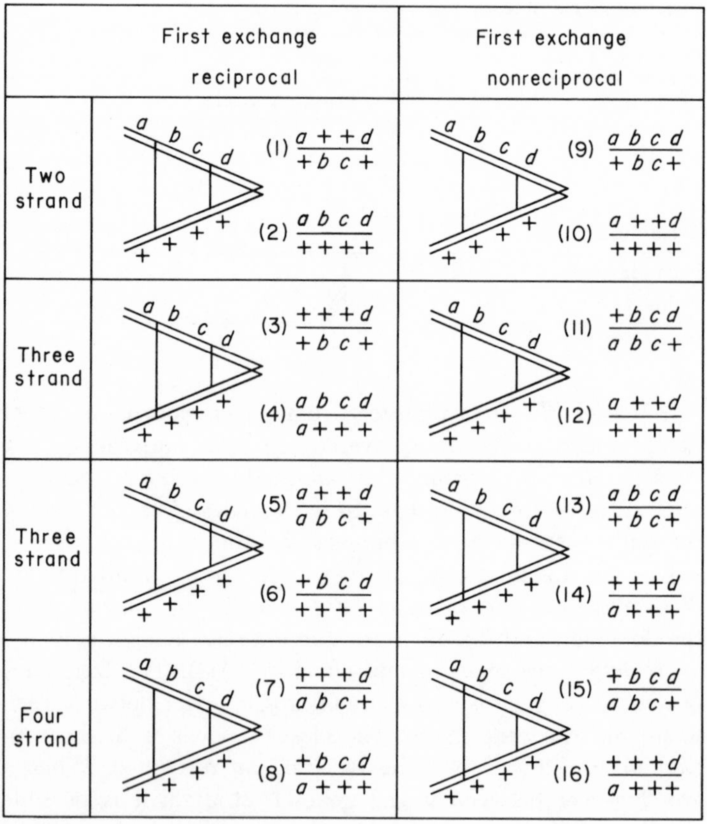

	First exchange reciprocal	First exchange nonreciprocal
Two strand	(1) $\frac{a + + d}{+ b c +}$ (2) $\frac{a b c d}{+ + + +}$	(9) $\frac{a b c d}{+ b c +}$ (10) $\frac{a + + d}{+ + + +}$
Three strand	(3) $\frac{+ + + d}{+ b c +}$ (4) $\frac{a b c d}{a + + +}$	(11) $\frac{+ b c d}{a b c +}$ (12) $\frac{a + + d}{+ + + +}$
Three strand	(5) $\frac{a + + d}{a b c +}$ (6) $\frac{+ b c d}{+ + + +}$	(13) $\frac{a b c d}{+ b c +}$ (14) $\frac{+ + + d}{a + + +}$
Four strand	(7) $\frac{+ + + d}{a b c +}$ (8) $\frac{+ b c d}{a + + +}$	(15) $\frac{+ b c d}{a b c +}$ (16) $\frac{+ + + d}{a + + +}$

(b)

Fig. 3–7. (*a*) The origin of homozygous eggs by crossing-over between the locus and the centromere in females of *Drosophila melanogaster* with occurs, not more than 16.7 percent homozygosis is obtained; when only reciprocal or nonreciprocal nonsister-strand crossing-over occurs, not more than 25 percent homozygosis is expected. By using a number of linked genes along the X chromosome, homozygosis for genes distal from the centromere did not exceed 25 percent. Furthermore, Welshons (1955) used an attached X chromosome to demonstrate that the frequencies of two-, three- and four-strand double crossing-over fit a ratio of 1:2:1 (Fig. 3-7*b*). The mutant flies with phenotypes resulting from a double crossing-over were analyzed by progeny testing to determine the genotype for each arm of the attached X chromosome. These results indicated that the chromatids involved in the first crossing-over were not subject to a bias in being involved in the second crossing-over.

CHROMOSOMES WITH UNUSUAL CENTROMERIC CHARACTERISTICS

The standard chromosome has a single centromere at a constant site usually identifiable as an achromatic segment or the site for the association between the chromosome and the spindle-apparatus fibers. While the evidence from morphological studies and from broken chromosomes indicates that the centromere is a complex structure, there is currently no accepted hypothesis or explanation to account for the origin of the centromere. A dicentric chromosome must cope with the

---

attached X chromosomes heterozygous for a sex-linked gene. (*b*) Crossover types recovered from the eight different double crossovers which can occur between nonsister chromatids of attached X chromosomes. The right exchange near the centromere may be reciprocal or nonreciprocal with equal frequency. Once this exchange is determined, the other exchange can occur in such a way as to form a two-, three-, or four-strand double exchange in the attached X chromosome. It is possible to recognize certain double exchanges by breeding the progeny of the females with the heterozygous attached X chromosome. Note that products of a reciprocal exchange are heterozygous to the left of the exchange site while those of a nonreciprocal exchange are homozygous. (*Welshons, 1955.*)

possibility that each centromere might go to opposite poles during mitosis or meiosis. Chromosomes lacking a specific site of centromeric activity are found in so few species that they are cytological curiosities. Finally, the appearance of sites with centromeric activity other than the centromere in standard chromosomes surely must be viewed as unusual.

Dicentric Chromosomes

A reciprocal translocation in which the centric segments fuse produces a dicentric translocation chromosome. While other mechanisms can yield dicentric chromosomes, these require a crossing-over in a heterozygous inversion. A transmissible dicentric chromosome was found in *Triticum aestivum* (Sears and Camara, 1952) and has provided information on the behavior of the centromeres during nuclear division. During the first meiotic division, the dicentric bivalent exhibits only a single site of association between one and the same centromere in each homolog and the spindle-apparatus fibers; a single dicentric chromosome has two active centromeres. In mitosis or the second meiotic division, both centromeres are active, but the dicentric chromosome usually proceeds toward one pole.

Multicentric Chromosomes

The horse threadworm (*Parascaris equorum*) has two chromosomal races, one with a single pair and the other with two pairs of chromosomes in the meiocytes, and each long chromosome has a median functional centromere. During early embryogenesis, however, the chromosomes fragment in cells which will yield somatic tissue producing a large number (50 to 72) of chromosomes; the terminal, heavily staining segments are lost. The cells developing into the germinal tissue and yielding meiocytes maintain the original chromosome number. One explanation for this unique situation assumes that the chromosomes are multicentric and that the median centromere is dominant in the pregerminal and germinal cells. In the cells committed to becoming somatic tissue, the chromosomes spontaneously break at predetermined sites so that each fragment has a single functional centromere and healed ends.

Diffuse Centromeric Activity

The chromosomes of species in the plant genus *Luzula,* in the scorpion *Tityus bahiensis,* and in homopteran and hemipteran species are associated with the spindle-apparatus fibers along their entire length at the appropriate stages of nuclear division. Detailed morphological studies have not provided any evidence for multicentric sites in these chromosomes. The discovery of chromosomes with diffuse centromeric activity in *L. purpurea* (Malheiros and de Castro, 1947) furnished the experimental material for extensive cytological studies. These unusual chromosomes go through the mitotic and meiotic divisions with no apparent difficulty and exhibit the usual characteristics of standard chromosomes, such as forming chiasmata. The chromosomes seem to be rather stiff and form an equatorial plane midway between the poles of the spindle apparatus at metaphase, and sister chromosomes are parallel as they separate during anaphase; the ends tend to lead as the chromosomes bend to conform to the shape of the spindle. When the chromosomes are broken by x-rays, the fragments retain their diffuse centromeric activity and are not lost during successive mitotic divisions, as generally happens to acentric fragments of standard chromosomes.

Neocentric Activity

The centromere is the only site of association between standard chromosomes and the spindle-apparatus fibers, indicating that no other site on the chromosome has this property. Neocentric activity has been reported for standard chromosomes in a few grass species and in *Lilium formosanum.* The most extensive investigations of this activity have been conducted in maize, but the situation may be unique in that it involves knobs on the standard chromosomes and an abnormal chromosome 10 (Rhoades and Vilkomerson, 1942). Heavily staining knobs are located at specific sites on different chromosomes. Varieties of maize differ with respect to the absence or presence of a knob at a specific site and the dimensions of the knob at a site. Chromosomes lacking a knob at a site do not generate one under any known conditions, and those with a knob do not lose it or exhibit marked differences in its size.

Neocentric activity in maize chromosomes is detected by the association of spindle fibers not only with the centromere but also with one or more sites on the same chromosome (Fig. 3-8). The sites of neocentric activity were positively identified as the knobs. The number of neocentric sites is determined by the number of knobs in the chromosomes in plants with the abnormal chromosome 10. In chromosomés with neocentric activity, one or both arms precede the centromere toward the pole during metaphase and anaphase in each meiotic division.

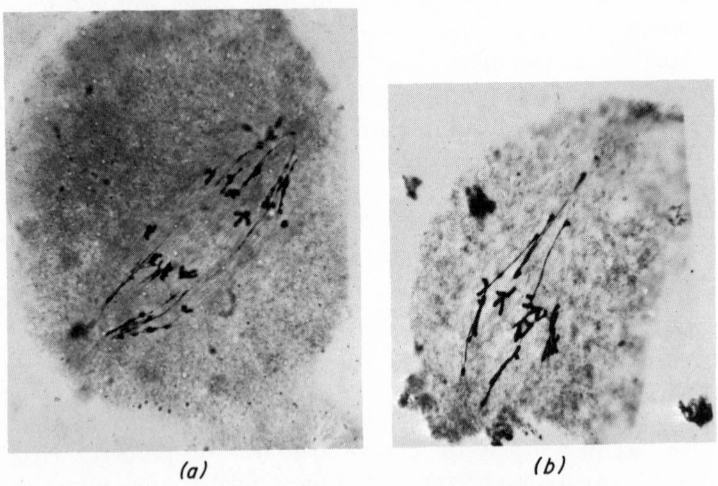

Fig. 3–8. Neocentric activity at (*a*) metaphase I and (*b*) metaphase II during microsporogenesis in maize. (*Photographs courtesy of Dr. M. M. Rhoades; Rhoades, 1955.*)

Neocentric activity is more pronounced in plants with a pair of abnormal chromosomes than in plants with one standard and one abnormal chromosome 10.

The centromere may produce a material which migrates along both arms of the chromosome and is responsible for the neocentric activity at the knob. To test this hypothesis, a heterozygous paracentric inversion was used to obtain an acentric fragment with a knob which had previously displayed neocentric activity in the standard chromosome. Although the knob in standard chromosomes displayed neocentric activity, the knob in the acentric fragment did not. The role of the abnormal chromosome 10 in neocentric activity, however, has not yet been clarified.

Genetic studies in plants homozygous for abnormal chromosome 10 and heterozygous for knobs and genic markers on the chromosomes with the knobs revealed significant discrepancies in the segregation of these markers (Rhoades, 1955). Crossing-over between the knobbed and knobless homologous chromosomes gives a knobbed and a knobless pair of chromatids in one chromosome. The knobbed chromatid with neocentric activity is closer to the pole than the knobless chromatid during anaphase I. Assuming that this situation is repeated during the second meiotic division, the knobbed chromatids preferentially enter the outer cells of the linear set of four megaspores during megasporogenesis. Since the basal or chalazal megaspore is generally the one to produce the female gametophyte, a knobbed rather than a knobless chromosome is more likely to be included in the egg nucleus. Genic markers close to the knob appear more frequently than expected, and the segregation data indicate a preferential segregation for these alleles which are not separated from the knob by crossing-over.

SUPERNUMERARY CHROMOSOMES

Plant and animal species are generally characterized by a specific and constant chromosome number. Many species, however, include individuals with one or more extra chromosomes which are not standard (or sex) chromosomes. These chromosomes are known as supernumerary or B chromosomes to distinguish them from the standard, or A, chromosomes and are not necessary for the proper functioning or viability of gametophytes or diploid individuals. While supernumerary chromosomes in plant or animal species can differ in length, they share a number of characteristics. Most of the cytogenetic information on these unusual chromosomes has come from plant species. Supernumerary chromosomes are usually shorter than the shortest standard chromosome in the species and are more or less completely heterochromatic.

Supernumerary chromosomes do not pair with the standard ones. Nonhomologous supernumerary chromosomes differing in length have been found in *Sorghum purpureo-sericeum* (Garber, 1950). Although a short and a long supernumerary chromosome do not pair, two short or two long chromosomes form a bivalent, indicating that the length of the chromosomes is not a factor in chiasma formation.

A single supernumerary chromosome usually has no effect on the phenotype, but their accumulation in one individual can lead to sterility in plant species. In *Plantago coronopus*, one B chromosome causes male but not female sterility. The most common phenotypic effect associated with increasing numbers of these chromosomes in one plant is sterility, but the particular number appears to be a species characteristic. Maize is unusual because one plant can have more than 20 supernumerary chromosomes. One explanation for this difference among species may involve a ratio between the diploid number and the supernumerary chromosomes. For example, diploid rye ($2n = 14$) has a maximum number of 10 B chromosomes and autotetraploid plants ($4n = 28$) a maximum number of 12 B chromosomes.

Reciprocal crosses between individuals with one or no B chromosome generally do not yield progeny with the expected number of these chromosomes. While the cross between a female parent with one B chromosome and a male with no

B chromosome usually yields progeny with one or no B chromosome, the progeny from the reciprocal cross generally includes individuals with zero, one, or two B chromosomes. In a few species, the progeny from the female and not the male will have more than the expected number of these chromosomes. Dhillon and Garber (1962) reported unexpected numbers of B chromosomes in progeny from reciprocal crosses in *Collinsia heterophylla* (Table 3-4). In species where the male with one B chromosome gives offspring with two B chromosomes, the increase results usually from a nondisjunction during the second mitotic division in the development of the male gametophyte from the microspore. One sperm nucleus receives two B chromosomes and the other no B chromosome, and the zygotic nucleus has either no or two B chromosomes. The frequency of progeny with two B chromosomes is a measure of the frequency of the mitotic nondisjunction, assuming no preferential fertilization of the egg nucleus by one of the sperm nuclei.

Attempts to formulate an acceptable hypothesis for the origin of supernumerary chromosomes generally have considered their heterochromatic appearance and apparent lack of Mendelian genes. In maize, the B chromosome at pachytene has a small euchromatic segment, suggesting that Mendelian genes might be present. A determined effort to detect alleles on this chromosome for genes on the standard chromosomes was unsuccessful (Randolph, 1941). Structural aberrations involving heterochromatic segments and either all or a functional portion of the associated centromere have offered a reasonable approach to constructing a supernumerary chromosome. Moens (1965) offered experimental evidence to support this explanation for the origin of such a chromosome. An isochromosome (2S·2S) was found in the progeny from a tomato plant with an extra chromosome 2 which has a heterochromatic short arm. The isochromosome with the iwo short heterochromatic arms did not pair with the standard chromosome, nor did two such isochromosomes form a bivalent. Two isochromosomes or three derivative smaller isochromosomes were the maximum

TABLE 3-4

Transmission of Supernumerary (B) Chromosomes to Progeny from *Collinsia sparsiflora* subsp. *arvensis* × *C. bruceae* (*a* × *br*) and from *C. solitaria*

Origin of Progeny	Number of Supernumerary Chromosomes				
	0	2	3	4	Total
			2 B: self-pollinated		
a × *br*	3	3	1	17	24
so	0	7	3	10	20
			2 B × 0 B		
(*a* × *br*) × *a*	1	14	0	8	23
(*a* × *br*) × *br*	3	6	0	0	9
so	4	16	1	0	21
			0 B × 2 B		
a × (*a* × *br*)	0	7	0	0	7
br × (*a* × *br*)	0	16	0	0	16
so	1	9	0	0	10
			3 B × 0 B		
so	4	10	0	14	28
			0 B × 3 B		
so	0	2	0	1	3
			4 B × 0 B		
(*a* × *br*) × *a*	2	13	0	9	24
(*a* × *br*) × *br*	0	1	0	2	3
so	6	12	1	26	45
			0 B × 4 B		
a × (*a* × *br*)	4	7	0	10	21
so	0	6	0	1	7

Source: After Dhillon and Garber, 1962.

numbers accumulated in one plant, and such plants did not exhibit any detectable phenotypic alterations, suggesting that the isochromosomes were genetically inert. While the 2S·2S isochromosome in tomato simulated a supernumerary chromosome, mitotic nondisjunction was not observed during the development of the male gametophyte.

Dhillon and Garber (1960) found an extra standard chromosome in the plant species *Collinsia heterophylla* which behaved like a pseudosupernumerary chromosome in the progeny from plants treated with colchicine. By appropriate self-pollinations, plants with up to five extra chromosomes were obtained, and they were indistinguishable from diploid individuals or from each other except for the sterility of plants with two or more chromosomes. Furthermore, the extra chromosomes apparently increased in number by a mitotic nondisjunction during the development of male or female gametophytes from spores. Finally, these extra chromosomes formed multivalents as well as bivalents and might yield an acceptable supernumerary chromosome, providing they did not pair with standard chromosomes and become heterochromatic.

REFERENCES

Beadle, G. W., and S. Emerson. 1935. Further studies on crossing over in attached X chromosomes in *Drosophila melanogaster.* Genetics 20:192–206.

Bridges, C. B. 1932. The genetics of sex in *Drosophila,* pp. 55–93. *In* W. C. Young (ed.), Sex and internal secretions. Williams & Wilkins, Baltimore.

Dhillon, T. S., and E. D. Garber. 1960. The genus *Collinsia.* X. Aneuploidy in *C. heterophylla.* Bot. Gaz. 121:125–133.

——— and ———. 1962. The genus *Collinsia.* XVI. Supernumerary chromosomes. Amer. J. Bot. 49:168–170.

Garber, E. D. 1950. Cytotaxonomic studies in the genus *Sorghum.* Univ. California Pub. Bot. 23:283–362.

Gowen, J. W. 1961. Genetic and cytological foundations for sex, pp. 3–75. *In* W. C. Young (ed.), Sex and internal secretions. Williams & Wilkins, Baltimore.

Khush, G. S., and C. M. Rick. 1968. Tomato telotrisomics: origin, identification, and use in linkage mapping. Cytologia 33:137–148.

Malheiros, N., and D. de Castro. 1947. Chromosome numbers and behavior in *Luzula purpurea* Link. Nature 160:156.

McClintock, B. 1938. The production of homozygous deficient tissues with mutant characteristics by means of the aberrant behavior of ring-shaped chromosomes. Genetics 23:315–376.

———. 1941. The association of mutants with homozygous deficiencies in *Zea mays*. Genetics 26:542–571.

Moens, P. B. 1965. The transmission of a heterochromatic isochromosome in *Lycopersicon esculentum*. Can. J. Genet. Cytol. 7:296–303.

Morgan, L. V. 1922. Non-criss-cross inheritance in *Drosophila melanogaster*. Biol. Bull. 42:267–274.

Randolph, L. F. 1941. Genetic characteristics of the B-chromosomes in maize. Genetics 26:608–631.

Rhoades, M. M. 1936. A cytogenetic study of a chromosome fragment in maize. Genetics 21:491–502.

———. 1955. The cytogenetics of maize, pp. 123–219. *In* G. F. Sprague (ed.), Corn and corn improvement. Academic, New York.

——— and H. Vilkomerson. 1942. On the anaphase movement of chromosomes. Proc. Nat. Acad. Sci. 28:433–436.

Schwartz, D. 1953*a*. The behavior of an X-ray induced ring chromosome in maize. Amer. Natur. 87:19–28.

———. 1953*b*. Evidence for sister-strand crossing over in maize. Genetics 38:251–260.

Sears, E. R., and C. A. Camara. 1952. A transmissible dicentric chromosome. Genetics 37:125–135.

Sen, N. K. 1952. Isochromosomes in tomato. Genetics 37:227–241.

Stern, C. 1936. Somatic crossing over and segregation in *Drosophila melanogaster*. Genetics 21: 625–730.

Welshons, W. J. 1955. A comparative study of crossing over in attached X-chromosomes of *Drosophila melanogaster*. Genetics 40:918–936.

SUPPLEMENTARY REFERENCES

Battaglia, E. 1964. Cytogenetics of B-chromosomes. Caryologia 17:245–299.

BROWN, S. W. 1954. Mitosis and meiosis in *Luzula campestris* DC. Univ. California Pub. Bot. 27:231–278.

DRONAMRAJU, K. R. 1965. The function of the Y chromosomes in man, animals, and plants. Advances Genet. 13:227–310.

MARKS, G. E. 1957. Telocentric chromosomes. Amer. Natur. 91:223–232.

NOVITSKI, E. 1963. The construction of new chromosomal types in *Drosophila melanogaster*, pp. 381–403. *In* W. J. Burdette (ed.), Methodology in basic genetics. Holden-Day, San Francisco.

RHOADES, M. M. 1952. Preferential segregation in maize, pp. 66–80. *In* Heterosis. Iowa State College Press, Ames.

SEARS, E. R. 1952. The behavior of isochromosomes and telocentrics in wheat. Chromosoma 4:551–562.

4 INTRACHROMOSOMAL STRUCTURAL ABERRATIONS

A standard chromosome has a characteristic allotment of chromosomal and genetic material oriented in a specific sequence. The reorientation, loss, or gain of a segment constitutes an intrachromosomal structural aberration. The breakage of standard chromosomes and the fusion of the raw ends of certain segments produce the primary aberrant chromosomes, and a cytogenetic study of the intrachromosomal structural aberrations depends on the transmissibility of such chromosomes. The reorientation of a chromosomal segment yields an inversion in which the sequence of the chromosomal and genic material is reversed only for the segment. The loss of a segment results in a deficiency and the addition of a segment in a duplication. The added segment may duplicate a segment already in the chromosome or come from another nonhomologous chromosome. The origin of these intrachromosomal structural aberrations as primary events can be explained by

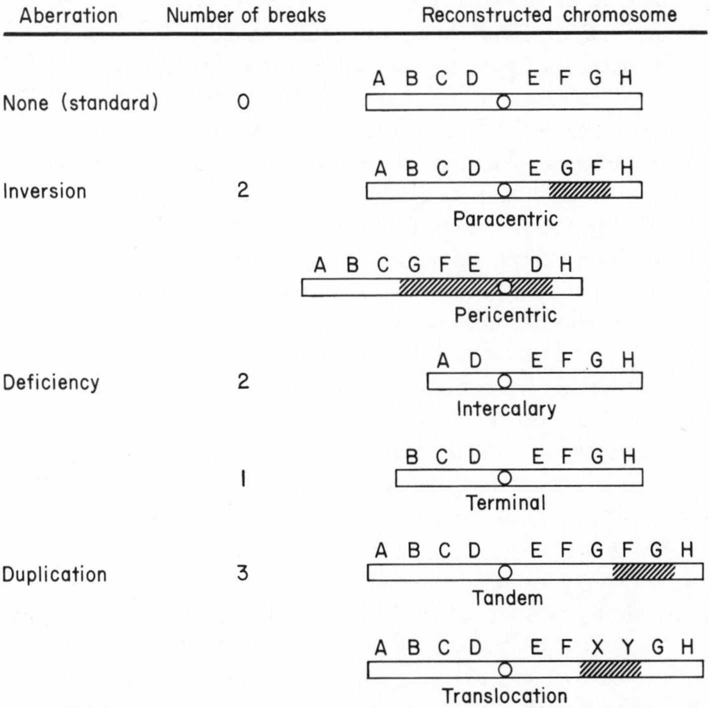

Aberration	Number of breaks	Reconstructed chromosome
None (standard)	0	A B C D E F G H
Inversion	2	A B C D E G F H Paracentric
		A B C G F E D H Pericentric
Deficiency	2	A D E F G H Intercalary
	1	B C D E F G H Terminal
Duplication	3	A B C D E F G F G H Tandem
		A B C D E F X Y G H Translocation

Fig. 4–1. Origin of primary intrachromosomal structural aberrations.

considering the number and sites of the breaks and the fusion of segments with raw ends (Fig. 4-1).

One break gives an acentric and a centric segment with a terminal deficiency. The replication of the terminally deficient chromosome yields two sister chromatids, each with a raw end, and the fusion of these ends gives a dicentric chromosome. The formation of a dicentric bridge during anaphase and the breakage of this bridge leads to a cycle of breakage, fusion, and bridge formation of which the end product is often a cell lacking the chromosome with the terminal deficiency. Consequently, a terminally deficient chromosome is not usually recovered in progeny unless delivered to a zygotic nucleus, where the raw end heals and subsequently behaves like a telomere. Two breaks in one chromosome lead to an inversion or to an intercalary deficiency, with the eventual loss of the acentric segment. Depending on the length and the site of the deficiency, the deficient chromosome can be responsible for the inviability or malfunctioning of the gametophytes, the zygote, or the developing embryo. Three breaks are needed to produce a duplication: two breaks in one chromosome produce a centric segment and two acentric segments, and one break in another chromosome furnishes the raw ends for the appropriate fusions. In this situation, the acentric segment with two raw ends fuses with each segment from the chromosome with a single break.

Intrachromosomal structural aberrations were first encountered in the course of genetic studies. For example, the effect of a heterozygous inversion on the recombination of certain linked genes in the chromosome was interpreted as a crossover-suppressor gene. For some deficiencies, the deleterious effect on viability was attributed to lethal genes, and in some species, a deficiency produced a phenotypic alteration which was interpreted as a genic mutation. The "genic" mutation responsible for Bar eye in *D. melanogaster* was later discovered to be a tandem duplication in the X chromosome. Once an intrachromosomal structural aberration was directly related to a specific genetic observation, it was possible to project from the genetic data to the pertinent aberrations. Although these aberrations might be minute and beyond cytological detection, cytogeneticists were sufficiently confident to interpret unusual genetic data in terms of a particular aberration. For example, a very small deficiency may not be visually detectable but can be inferred by pseudodominance, that is, the loss of a small segment with the dominant allele responsible for a recessive phenotype in presumably heterozygous individuals. The prokaryote geneticist frequently indulges in such projections, particularly when the breeding data call for deficiencies.

INVERSION

Although inversions have been reported in many plant and animal species, their frequency is relatively low and usually restricted to an occasional individual or population in the species. Ionizing radiation or radiomimetic chemicals have been successfully used to obtain inversions because these agents produce chromosomal breaks. The classification of an inversion is related not to the length of the reoriented segment but to the presence (pericentric) or absence (paracentric) of the centromere within the segment. Depending on the site and length of the inversion, the morphology of the standard chromosome can be obviously altered in its arm ratio by a pericentric inversion, while both types of inversion can displace chromosomal markers, such as heterochromatic segments, from their standard location.

Single Crossing-over in Heterozygous Inversions

The pairing of homologous segments at pachytene in an inversion heterozygote is responsible for a loop which includes the inverted segment. The inversion loop is almost diagrammatic in the salivary-gland chromosomes of *Drosophila* and in species with excellent pachytene chromosomes. When the segment is short, a loop is not likely to be formed because the homologous inverted and standard segments do not synapse. The pairing of a relatively long inverted and standard segment can yield a bivalent without a loop and with terminal segments which cannot pair because of nonhomology. The site and length of the inversion loop and the presence or absence of the centromere within the loop are readily determined in species with good pachytene chromosomes (Fig. 4-2).

The cytological consequences of a single crossing-over within a heterozygous inversion provide a visual basis for distinguishing between a paracentric and a pericentric inversion and accounting for the sterility caused by either type of inversion. A dicentric chromatid bridge and acentric fragment at

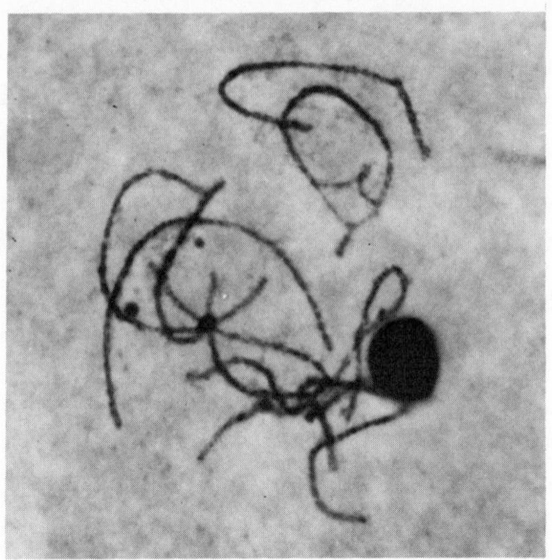

Fig. 4–2. Heterozygous pericentric inversion in chromosome 2 of maize. (*Photograph courtesy of Dr. M. M. Rhoades.*)

anaphase I characterize a single crossing-over in the paracentric but not the pericentric inversion (Fig. 4-3). When the pachytene chromosomes cannot be examined to detect the inversion loop, the bridge and fragment are acceptable evidence for a heterozygous paracentric inversion. The products of a single crossing-over in either type of inversion are two chromosomes with one deficient chromatid and a standard or an inversion chromatid. At the end of meiosis, two of the four haploid nuclei will have a deficient chromosome.

In plants, spores with deficient chromosomes are likely to abort and not develop into viable gametophytes. When the fertility of plants heterozygous for a paracentric or pericentric inversion is compared, the percentage of aborted pollen grains is generally correlated with the frequency of crossing-over in the inverted segment. This correlation provides a valuable tool for investigating factors which influence crossing-over without

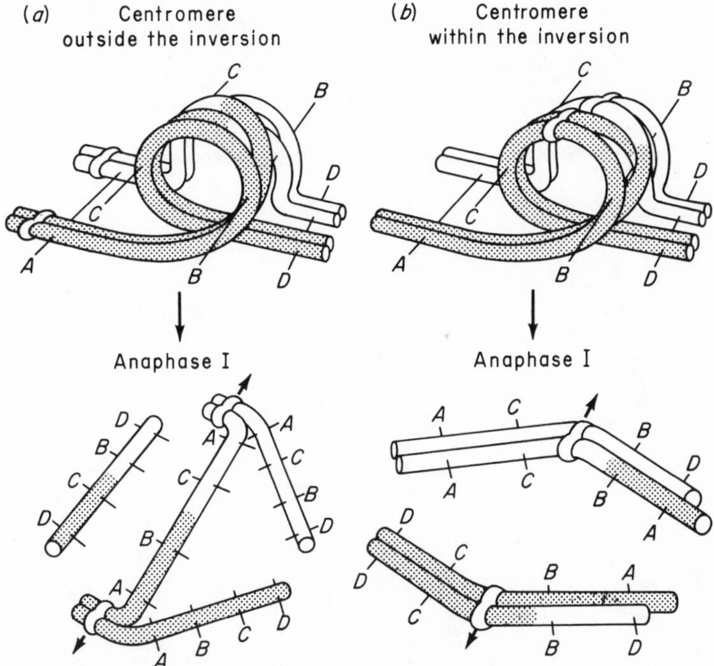

Fig. 4–3. The anaphase I configurations resulting from a crossover within a paracentric (*a*) or a pericentric (*b*) inversion. The dicentric chromatid and acentric fragment at anaphase I constitute cytological evidence for a heterozygous paracentric inversion. (*From Adrian M. Srb, Ray D. Owen, and Robert S. Edgar. General genetics, 2d ed. W. H. Freeman and Company. Copyright © 1965.*)

resorting to linked genes and breeding data. The percentage of aborted ovules is determined by using the inversion heterozygote as the female parent and comparing the seed set with that for a standard female parent. The percentage of aborted ovules is directly related to the frequency of single crossing-over in a pericentric inversion but not in a paracentric inversion. In animal species, deficient gametes but not the deficient zygotes usually function, so that the lethal effect of the defi-

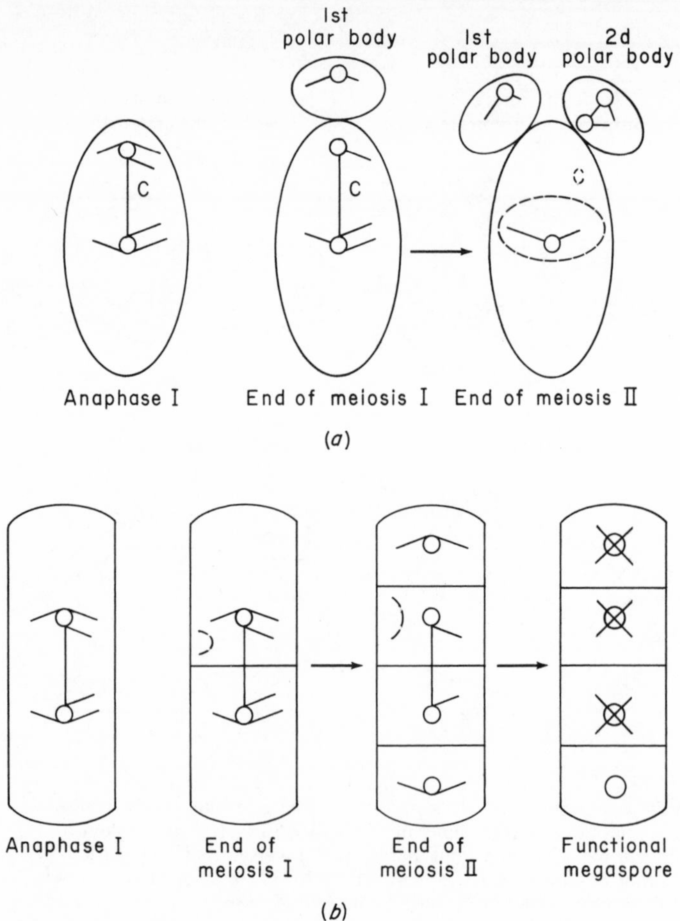

Fig. 4–4. Diagrams accounting for the fertility of females heterozygous for a paracentric inversion. (*a*) Meiosis is initiated when the nucleus is eccentric, so that the first polar body receives an intact standard or inversion chromatid at the end of the first meiotic division. During the second meiotic division, the dicentric chromatid presumably is included in the second polar body, and the egg nucleus contains either an intact standard or inversion chromosome. The acentric fragment associated with the dicentric chromatid bridge presumably disintegrates in the egg cytoplasm. (*b*) At the end of the first meiotic division, the orientation of the dicentric chromatid presumably is responsible for the position of the intact standard or inversion chromatid so that the terminal cells of

ciency has to be established for reciprocal crosses, noting the fertility of the male or female. Males heterozygous for a pericentric or a paracentric inversion and the females with the pericentric inversion exhibit reduced fertility.

The absence of an obvious sterility for females heterozygous for a paracentric inversion results from the directed orientation of the chromosomes in the dicentric chromatid bridge (Fig. 4-4). At anaphase I, the intact chromatids at each pole are so oriented that they will be included in the outer megaspores and the deficient chromatids in the inner megaspores. Consequently, the basal megaspore receives either a standard or an inversion chromosome and the gametophyte is not deficient. In oogenesis, the orientation is responsible for the inclusion of the deficient chromosomes in two polar nuclei and of an intact or an inversion chromosome in the egg nucleus.

The heterozygous inversion acts as a crossover suppressor for linked genes within the inverted segment or relatively close to this segment. While a single crossing-over in the segment produces recombinants, they are included in the deficient spores or zygotes which do not contribute to the scorable progeny. Muller (1928) first exploited this characteristic of a heterozygous inversion by placing an inverted segment between the loci for a lethal and the Bar mutation in *D. melanogaster* in constructing the *ClB* chromosome to detect x-ray-induced mutations. The Bar phenotype served as a visible marker to ensure that the chromosome also had the lethal gene; the inversion between the markers eliminated recombinants in the progeny from the cross involving the *ClB* females. It is interesting to note that the lethal "gene" was a deficiency, the crossover-suppressor "gene" an inversion, and the Bar "gene" a duplication.

---

the linear quartet of megaspores will receive a standard or inversion chromosome. One terminal megaspore develops into the female gametophyte.

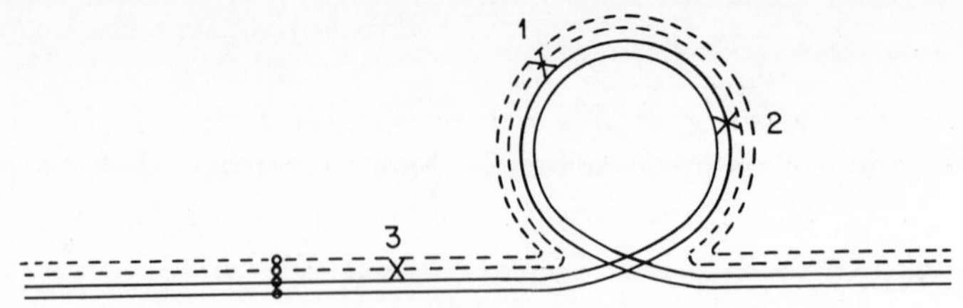

Sites of crossing over

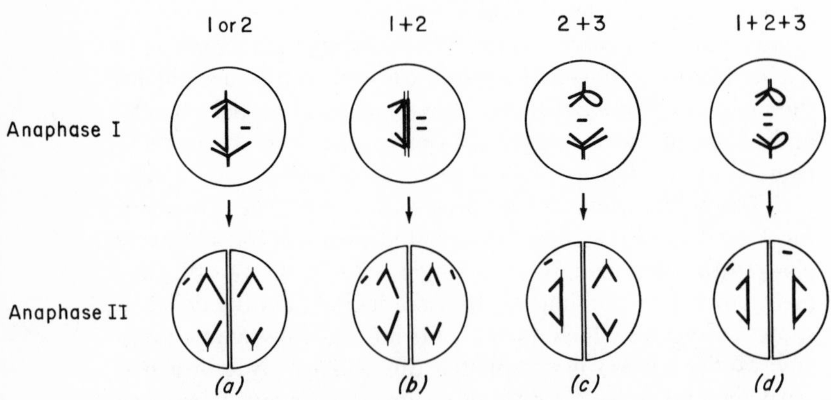

Fig. 4–5. The origin of different types of anaphase I and II configurations resulting from crossovers at one or more sites (*numbered*) within the heterozygous paracentric inverted segment and the intercalary segment between the centromeres (*clear circles*) and the inversion. Note the strands involved in each crossover at the different sites. (*a*) Single crossover within the inverted segment. (*b*) Four-strand double crossover within the inverted segment. (*c*) Three-strand double crossover, one in the inverted and the other in the intercalary segment. (*d*) Triple crossover, two in the inverted and the third in the intercalary segment. The anaphase I configurations distinguish (*a*) from (*b*) and both from (*c*) and (*d*), while the anaphase II configurations distinguish (*c*) from (*d*) and both from (*a*) and (*b*). The acentric fragment or fragments may not be detected in all cells. (*After McClintock, 1938a.*)

Double Crossing-over in a Paracentric Inversion

McClintock (1938*a*) found a paracentric inversion in chromosome 9 of maize which was sufficiently long to allow double crossing-over within the inverted segment and far enough from the centromere to allow a single crossing-over in the intercalary segment between the centromere and the inversion (Fig. 4-5). The different chromosomal configurations in the pollen mother cells at anaphase I and II were related to a single or a double crossing-over in the inverted segment and to a single crossing-over in the intercalary segment and in the inverted segment (Fig. 4-5). Three configurations were unique and resulted from specific combinations of crossing-over: (1) a four-strand double crossing-over in the inverted segment, (2) a three-strand double crossing-over, one in the inverted and the other in the intercalary segment, and (3) a triple crossing-over, one in the intercalary segment and a four-strand double crossing-over in the inverted segment. The first combination yields two acentric fragments and a dicentric double chromatid bridge at anaphase I (Fig. 4-6*b*); the second com-

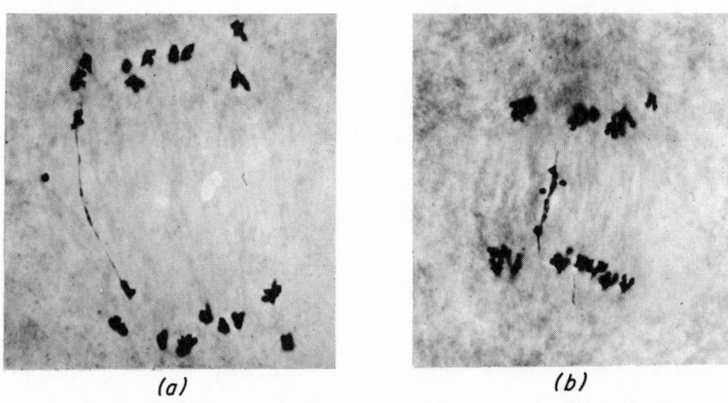

Fig. 4–6. Anaphase I products of crossing-over in a heterozygous paracentric inversion in maize. (*a*) Dicentric chromatid bridge and acentric fragment resulting from a single crossover, (*b*) Double dicentric chromatid bridge and two acentric fragments resulting from a four-strand double crossover. (*Photographs courtesy of Dr. M. M. Rhoades; Rhoades, 1955*).

bination produces a single acentric fragment and a "ring" chromosome at anaphase I which is responsible for a dicentric bridge in one of the two secondary meiocytes at anaphase II; and the third combination gives two acentric fragments and two "ring" chromosomes at anaphase I which are responsible for a dicentric bridge in each of the two secondary meiocytes at anaphase II. While a three-strand double crossing-over yields a "ring" chromosome, rod chromosome, and acentric fragment at anaphase I, the "ring" chromosome and acentric fragment are not always detected. The frequencies of the configurations at anaphase I and II are summarized in Table 4-1. Assuming that the three types of double crossing-over occurred at random (25 percent two-strand:50 percent three-strand:25 percent four-strand) and interference was negligible, the frequencies of the unique products of double crossing-over or of the triple crossing-over can be used to calculate the frequency of each type of double crossing-over and of a single crossing-over in the intercalary segment.

Breakage-Fusion-Bridge Cycle in Maize

The breakage of the dicentric chromatid bridge at anaphase I produces terminally deficient chromosomes with raw ends at the conclusion of meiosis. Although a deficient microspore does not develop into a viable gametophyte when the deficiency is relatively long, the chance inclusion of the compensating acentric fragment in the microspore nucleus permits normal development. During each of the two mitotic divisions from microspore to male gametophyte, the broken chromosome replicates, and the sister chromatids fuse at the raw ends. The resulting dicentric chromosome forms a bridge at each anaphase, which breaks, yielding two terminally deficient chromosomes. Because the genetic content of the male gametophyte allows the two deficient nuclei to function in the double fertilization, the zygotic nucleus and the primary endosperm nucleus each have a terminally deficient chromosome with a raw end. The deficient zygote produces a normal embryo when

TABLE 4-1

Configurations at Anaphase I and II in Maize Plants Heterozygous for a Paracentric Inversion

A. Anaphase I in Pollen Mother Cells

Plant	No Bridge; No Fragment	Single Bridge; One Fragment	Double Bridge; Two Fragments	No Bridge; One Fragment	% Single Bridge; One Fragment	% Double Bridge; Two Fragments	% No Bridge; One Fragment
1013-4	34	30	1	3	44.1	1.4	4.4
	74	54	4	4	39.7	2.9	2.9
1013-9	10	6	1	1	33.3	5.5	5.5
	18	19	0	0	51.3	0.0	0.0
	43	40	4	4	43.9	4.4	4.4
	16	16	1	3	44.4	2.7	8.3
Total	195	165	11	15	42.7	2.8	3.8

B. Anaphase II, Considering the Two Secondary Meiocytes as One Unit

Plant	No Bridge; No Fragment	No Bridge; Fragment in Cytoplasm of One Cell	No Bridge; Fragment in Spindle of One Cell	No Bridge; Two Fragments in Cytoplasm	Bridge in One Cell; Fragment in One Cell	% No Bridge; One Fragment	% No Bridge; Two Fragments	% One Bridge; One Fragment
1013-11	36	25	14	1	1	50.6	1.3	1.3
	51	25	20	0	3	45.4	0.0	3.0
	25	20	7	0	2	50.0	0.0	3.7
1013-13	93	32	23	4	6	34.8	2.5	3.8
Total	205	102	64	5	12	42.7	1.3	3.0

Source: After McClintock, 1938*a*.

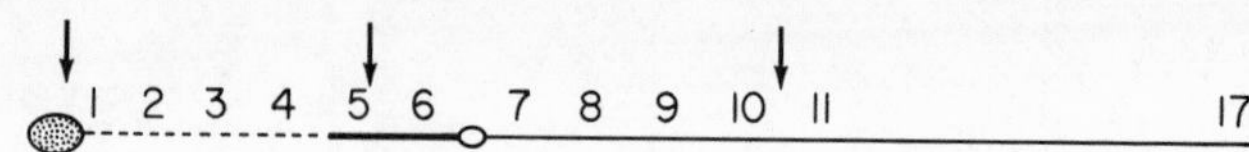
1 2 3 4 5 6 7 8 9 10 11
17

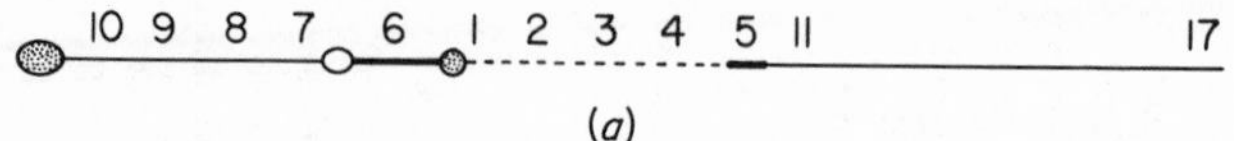
10 9 8 7 6 1 2 3 4 5 11
17

(*a*)

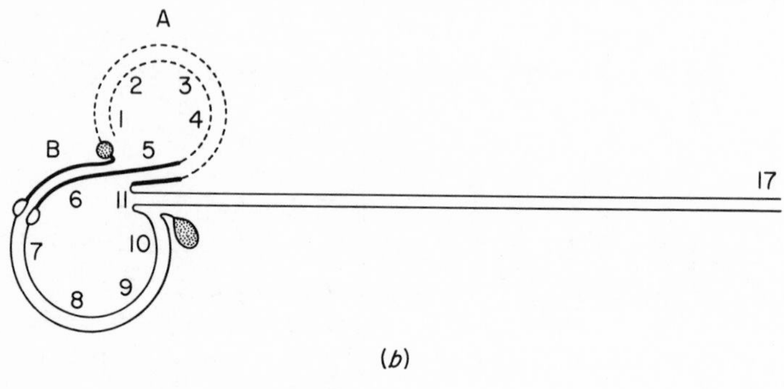
A
2 3
1 4
B 5
6 11
17
7 10
8 9

(*b*)

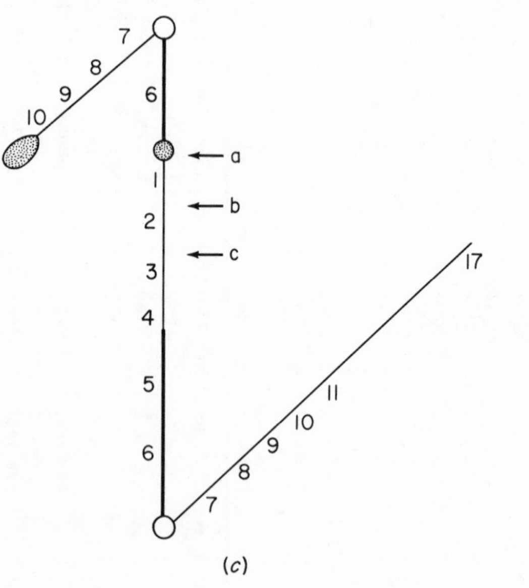
7
8
9
10
6
a
1
b
2
c
3
4
5
6
17
11
10
9
8
7

(*c*)

the deficiency does not interfere with normal embryogenesis. The triploid primary endosperm nucleus compensates for the deficiency, and the numerous mitotic divisions eventually produce the endosperm tissue. Although this method can yield viable zygotes with a terminally deficient chromosome, the chance inclusion of the compensating acentric fragment in the deficient microspore nucleus and the extent of the deficiency in the zygotic nucleus are responsible for the very low number of viable zygotes.

McClintock (1941) devised two different methods of obtaining terminally deficient chromosomes in maize which did not require a compensating acentric fragment to get functional male gametophytes. One method used a chromosome with a tandem duplication in which one segment was inverted, and the second method used a double included inversion. In either method, a crossing-over in the appropriate chromosome segment yields a dicentric chromatid bridge but no acentric fragment at anaphase I. The breakage of the bridge in different pollen mother cells can produce different terminally deficient chromosomes with a raw end. Considering the great number of pollen mother cells produced by maize plants, a reasonable

---

Fig. 4–7. (*a*) *Top:* Diagram of a standard chromosome 9 of maize with a large terminal knob (*stippled*) on the short arm, widely spaced small chromomeres in the distal two-thirds of the short arm (*broken line*), a proximal deep-staining segment in the short arm (*wide line*), centromere (*clear circle*), and long arm (*narrow line*). (*a*) *Bottom:* Diagram of the reconstructed chromosome 9 resulting from a break (*arrow*) in the terminal knob, in the deep-staining segment and in the long arm. (*McClintock, 1941.*) (*b*) Synaptic configuration for a standard chromosome 9 without the terminal knob and the reconstructed chromosome 9. Region A includes segments 1 to 5, region B segment 6 (between the small knob and the centromere), and region C segments 7 to 10. (*McClintock, 1941.*) (*c*) Diagram of how the dicentric chromatid bridge at anaphase I produces a crossover in region A. The intercentromeric segment includes all the genes of the short arm of chromosome 9 plus a duplication of the proximal part of the knob and the proximal fourth of the short arm. The arrows indicate potential sites of breakage yielding chromosomes with no deficiency or with a terminal deficiency. (*Rhoades, 1955.*)

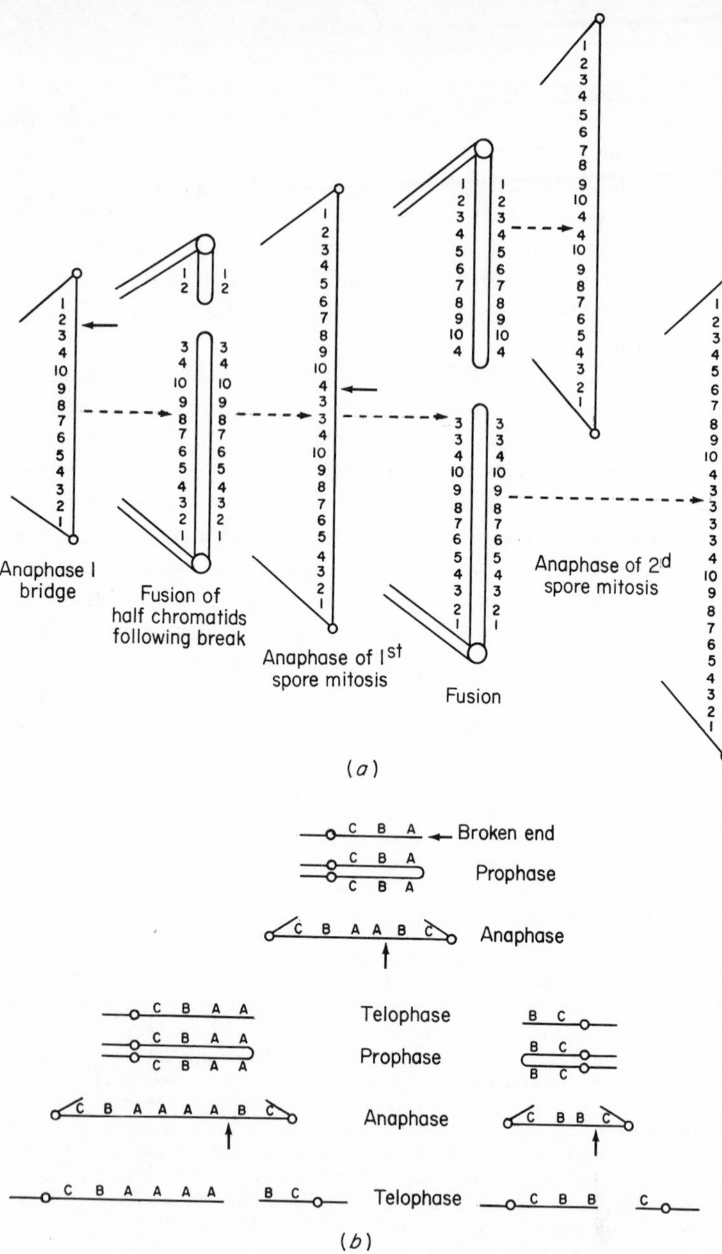

Anaphase I bridge
Fusion of half chromatids following break
Anaphase of 1st spore mitosis
Fusion
Anaphase of 2d spore mitosis
(a)
Broken end
Prophase
Anaphase
Telophase
Prophase
Anaphase
Telophase
(b)

number of functional pollen grains can be expected to include such chromosomes.

Pollen grains from a plant with the dominant alleles for four linked genes (*yg-2-c-sh-wx*) in the short arm of chromosome 9 with a large terminal knob were irradiated with x-rays and then used to pollinate plants homozygous for the recessive alleles. The hybrid seeds were colored (*C c c*), nonshrunken (*Sh sh sh*), and nonwaxy (*Wx wx wx*). Hybrid plants were crossed to homozygous recessive plants with a chromosome 9 lacking a terminal knob. Most of the seeds from one hybrid were uniformly colored, but a few were variegated. The hybrid parent had a reconstructed chromosome and a standard chromosome 9. The reconstructed chromosome resulted from one break in the terminal knob, a second break in the heterochromatic segment in this arm and adjacent to the centromere, and a third break approximately in the middle of the long arm (Fig. 4-7*a*). In the hybrid plant, the synapsis of the homologous segments produced two inversion loops at pachytene (Fig. 4-7*b*). In meiocytes with a crossing-over in one loop

Fig. 4–8. (*a*) Diagram of the breakage-fusion-bridge cycle in male gametogenesis following the formation of a dicentric chromatid bridge at anaphase I by crossing-over within at heterozygous paracentric inversion. The solid arrow indicates the site of breakage; the numbers represent the individual chromosomal segments. (*Rhoades, 1955; from McClintock, 1941.*) (*b*) A representative illustration of the method by which variegation may be produced in tissues carrying a chromosome with a broken end. The clear circle represents the centromere. The dominant genes *A, B,* and *C* are carried by the arm with the broken end, *A* being near the broken end and *C* near the centromere. The homolog of this chromosome (not diagrammed) is considered to be normal and to carry the genes *a, b,* and *c*. Division of this broken chromosome results in fusion at the position of breakage between the two split halves (*prophase, second diagram from top*). This is followed by a bridge configuration in the following anaphase (*anaphase, third diagram from top*). The arrow points to the position of breakage, the two broken chromosomes entering the sister telophase nuclei (*telophase, right and left*). This process is repeated in successive divisions. One such division is diagrammed below each of these two telophase chromatids. The diagrams illustrate how dominant genes may be deleted or reduplicated following the breakage-fusion-bridge cycle. (*McClintock, 1941.*)

(region A), the chromatid bridge has all the genes of the short arm (standard chromosome) and a duplication for the proximal part of the knob and the proximal one-quarter of the short arm (Fig. 4-7*c*).

A break in the bridge during anaphase I at site *a* yields a chromatid with all the genic material but lacking the unessential knob. A break at site *b* gives a chromatid deficient for segment 1. Finally, a break at site *c* produces a chromatid missing segments 1 and 2. These three types of chromatids have a raw end which will be involved in a breakage-fusion-bridge cycle (Fig. 4-8) during the two mitotic divisions from microspore to male gametophyte or during the three mitotic divisions from megaspore to female gametophyte. The sperm or egg nucleus, therefore, can have a chromosome 9 with a raw end and a duplication or deficiency for segments of different lengths, depending on the sites of breakage in the mitotic divisions. Furthermore, some chromosomes can be deficient for very small terminal segments of the short arm. Although the chromosome with the raw end is involved in a breakage-fusion-bridge cycle at each mitotic division in the development of the endosperm, the raw end of this chromosome "heals" in the zygotic nucleus. The variegated seeds have embryos with the chromosome 9 involved in the breakage-fusion-bridge cycle in the endosperm. The cytogenetic analysis of plants from such embryos will be presented in the discussion of deficiencies.

Multiple Inversions

A population may include members homozygous for a single inversion, the standard sequence, or heterozygous for the inversion. Spontaneous chromosomal breaks can produce inversions for different segments of one chromosome of one or different individuals. Consequently, one chromosome can have two inverted segments as products of two events separated in time. The physical relation between these inversions is the basis for classifying double inversions by the types of inversion loops observed in the heterozygotes. The salivary-gland chromosomes of *Drosophila* are particularly useful for this type of cytological examination because of their exquisite detail, but good pachytene chromosomes serve the same purpose. A heterozygote with two independent inversions displays two loops separated by a chromosomal segment; one with two included inversions yields two loops in the form of a figure eight to one side of the chromosomes; and the third type with two overlapping inversions also gives two loops but on opposite sides

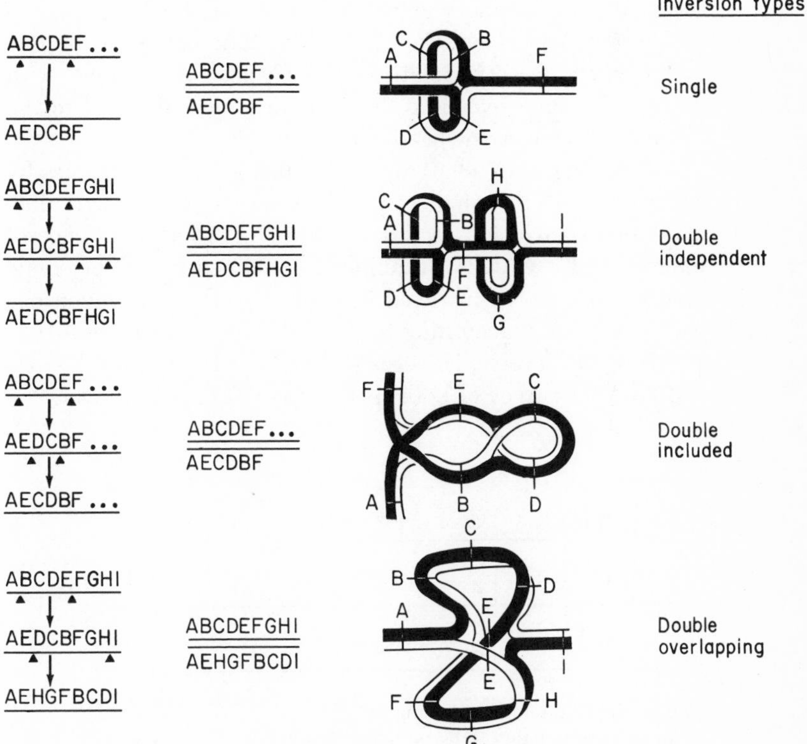

Fig. 4–9. Diagrams of pairing for the salivary-gland chromosomes in *Drosophila pseudoobscura* heterozygous for different types of inversions. Configurations at the left indicate the origin of the inversions; the black triangles indicate the breakage points. (*Dobzhansky, 1951.*)

of the chromosomes (Fig. 4-9). The double overlapping inversions in *D. pseudoobscura* have been useful tools for evolutionary studies in this species (Sturtevant and Dobzhansky, 1936), which were later extended to the related species *D. persimilis*.

A survey of populations of *D. pseudoobscura* revealed heterozygous single or double overlapping inversions in chromosome III, and their location and length were accurately determined in the salivary-gland chromosomes. Assuming a standard sequence for this chromosome, each inversion was assigned to a specific segment. For example, the single inversion in the Arrowhead population was different from the one in the Pike's Peak population. The Cochise population, with two overlapping inversions, had one inversion identical to the inversion in the Arrowhead population.

A phylogenetic chart was constructed for both species by assuming that the single inversions involved in the double overlapping inversions had occurred in an orderly sequence of independent events. For example, two sequences for the single inversions A and B involved in double overlapping inversions can be proposed according to the following logic (Fig. 4-10): I → II or the reverse, II → I and then II → III, but not I → III or III → I. A survey of both species gave 27 different inversions

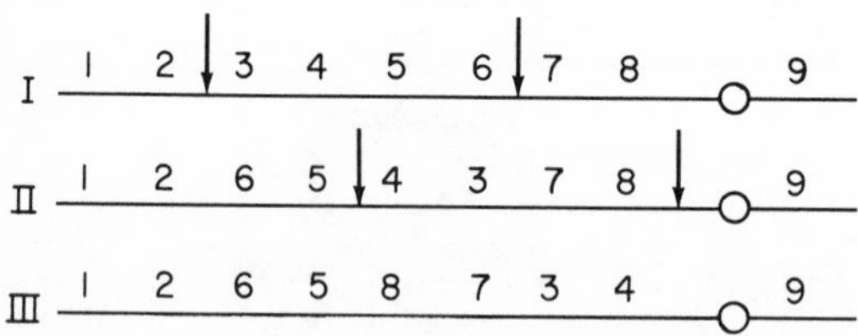

Fig. 4–10. Diagrams illustrating three chromosomes assumed to be sequentially related to each other by overlapping inversions. The arrows indicate the site of breaks resulting in each inversion. The order of the sequences must be I→II→III, III→II→I, or I←II→III.

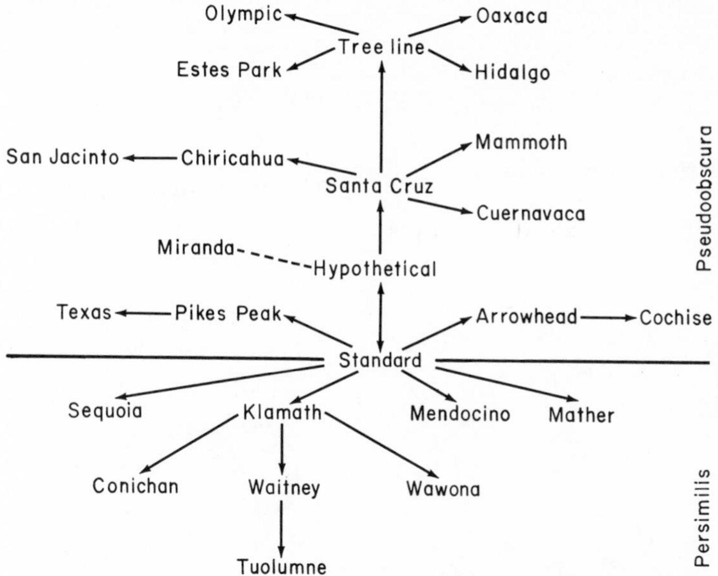

Fig. 4–11. Phylogenetic chart of different gene sequences found in the third chromosome of *Drosophila pseudoobscura* and *D. persimilis* by the use of overlapping inversions. Sequences joined by a single arrow differ by a single paracentric inversion. Sequences above standard are found only in *D. pseudoobscura,* and those below standard only in *D. persimilis;* standard was found in both species. (*Dobzhansky, 1951.*)

in chromosome III which could be sequenced to yield an acceptable phylogenetic chart (Fig. 4-11). One inversion needed to bridge a gap in the sequence from the standard chromosome to the Santa Cruz population, however, has not been found in *D. pseudoobscura.* This hypothetical inversion was present in a related species *D. miranda.* Assuming a common standard chromosome in both *D. pseudoobscura* and *D. persimilis,* the former species had one group of inversions and the latter species had the second group of inversions. This ingenious use of chromosomal polymorphism to solve phylogenetic problems will be encountered again in the discussion of reciprocal translocations in *Oenothera.*

DEFICIENCY

A missing segment constitutes a deficiency for a standard chromosome. The site and length of a deficiency often can be determined by examining pachytene chromosomes in a deficiency heterozygote (Fig. 4-12). The homologous segment in the standard chromosomes forms at the site of the deficiency a loop, or buckle, the length of which is a measure of the length

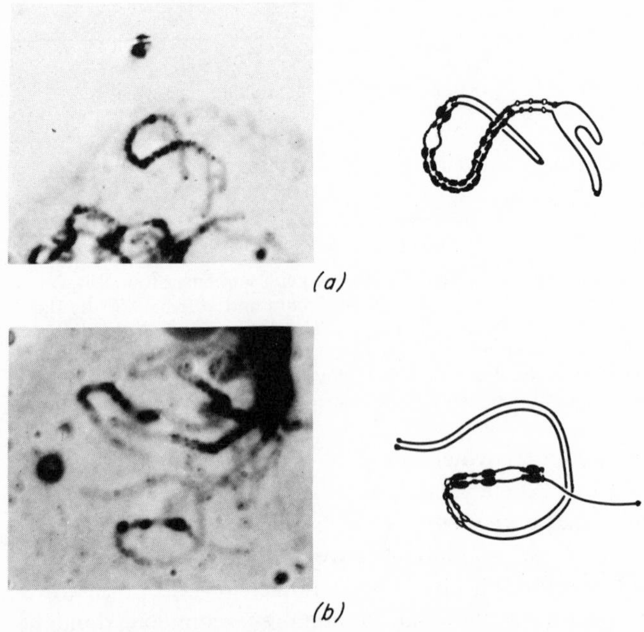

Fig. 4–12. Heterozygous deficiency configurations at pachytene in tomato. (*Photographs courtesy of Dr. C. M. Rick*). (*a*) Intercalary deficiency entirely within the euchromatin of the long arm of chromosome 4. The deficient region extends to and includes the prominent knob, and that part of the euchromatin is nonhomologously paired in a loop-back fashion. (*Khush and Rick, 1967.*) (*b*) Terminal deficiency in the short arm of chromosome 6. (*Khush and Rick, 1968.*)

of the deficiency. The detailed banding of the salivary-gland chromosomes in *Drosophila* permits detection of deficiencies so minute that equivalent deficiencies cannot be detected in pachytene chromosomes (Fig. 4-13). When a relatively long deficiency is near the end of the chromosome, the standard chromosome usually appears to be longer than the deficient chromosome in the heterozygote. Some species include individuals with heteromorphic bivalents (homologous chromosomes of different length) which might have resulted from a subterminal deficiency in one chromosome.

While gametophytes with a deficient chromosome are often not viable, the female gametophyte is more likely to tolerate a deficiency than the male gametophyte. One explana-

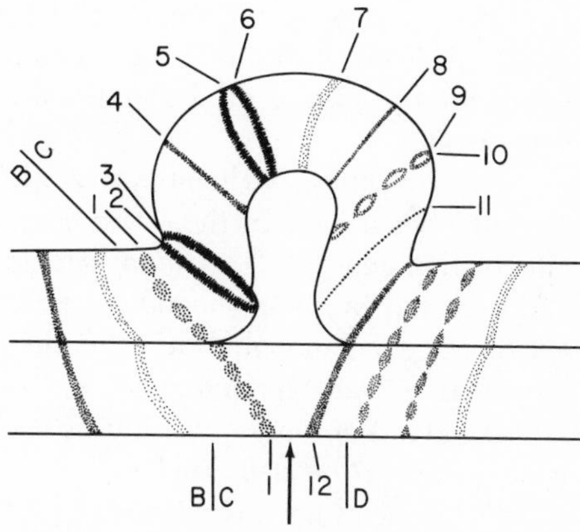

Fig. 4–13. The pairing in a segment of the salivary gland X chromosomes in *Drosophila melanogaster* heterozygous for a deficiency responsible for the Notch phenotype. The lower deficient chromosome lacks the segment containing bands 3C2 through 3C11. (*From Curt Stern. Principles of human genetics, 2d ed. W. H. Freeman and Company. Copyright © 1960.*)

tion for this difference is related to the specialized function of the male gametophyte, which has to produce a pollen tube to effect fertilization. Deficiencies may interfere not only with the normal development of the gametophyte, to produce an aborted pollen grain, but also with the normal development of the pollen tube. Long deficiencies are more frequently lethal than short deficiencies. In one species, deficiencies of approximately equal length can be lethal when the loss occurs in certain regions of chromosomes or nonlethal in other regions of chromosomes. These observations suggest that the genes for viability or function are not randomly distributed throughout one or all the chromosomes.

The deficient gametes of animals function, and heterozygous zygotes yield viable individuals. In *D. melanogaster*, certain deficiencies in the Y chromosome are responsible for nonmotile but apparently viable sperm, and in these cases, the deficiencies may be viewed as operationally lethal. Homozygous deficiencies in plants or animals, however, are generally lethal for the zygote or the embryo at some stage of its development.

Heterozygous or homozygous deficiencies can produce mutant phenotypes. In the latter case, a deficiency will simulate a lethal gene. In *D. melanogaster*, certain deficiencies were first detected by their phenotypic effect, and because their inheritance gave a monohybrid ratio, they were treated as though they were indeed Mendelian genes. By carefully examining the salivary-gland chromosomes in this species, many lethal genes were shown to be deficiencies.

Deficiencies have furnished the cytogeneticist with a valuable tool for assigning genes to specific regions of the chromosome. Although deficiencies arise spontaneously, deficiencies for mapping purposes are obtained by the breakage of chromosomes of pollen grains or sperm with ionizing radiation in an individual homozygous for dominant alleles and fertilizing homozygous recessive females. The hybrid progeny will include mostly progeny with the dominant allele and occasional individuals with a recessive phenotype. When the genes are linked, the recessive progeny can have one or more phenotypes determined by very closely linked mutant genes. These unexpected progeny are assumed to be hemizygous for the locus, and the expression of the recessive phenotypes is termed *pseudodominance.* Although the possibility of an induced mutation must be considered, pseudodominance is usually assumed to indicate the loss of the chromosomal segment carrying the dominant allele. A cytological study of pachytene chromosomes or of salivary-gland chromosomes in the heterozygote often reveals the deficiency buckle. In *D. melanogaster,* it has been possible to detect deficient segments with a resolution far exceeding any plant species or other animal species.

The precise localization of a genic locus requires a number of independent deficiencies, detected first genetically by pseudodominance. Such a method was used in *D. melanogaster* for deficiencies of the white-eye locus in the X chromosome. The salivary-gland chromosomes of the heterozygotes revealed that the deficiencies overlapped, and in a sample of 14 independent deficiencies, one missing band (3C2) was common (Fig. 4-14). Consequently, the locus for white eye is either in or very close to this band in the salivary chromosomes. The method has been extensively used in *D. melanogaster* and has been applied to a number of eukaryotic and prokaryotic species.

Overlapping deficiencies were used to sequence the mutant sites in the *r*II locus of coliphage T4 (Benzer, 1961). In this strain, *r*II mutants arise spontaneously as genic mutations or as deficiencies. Mutants reverting to the wild-type phenotype are presumably genic, and those failing to yield revertants are presumably deficiencies. When the deficiency mutants are crossed and do not yield wild types by recombination, it is reasonable to assume that the deficiencies shared a common missing segment. This ingenious adaptation of the method using overlapping deficiencies developed for *D. melanogaster* was successfully applied to mapping mutant sites in the *r*II locus of coliphage T4.

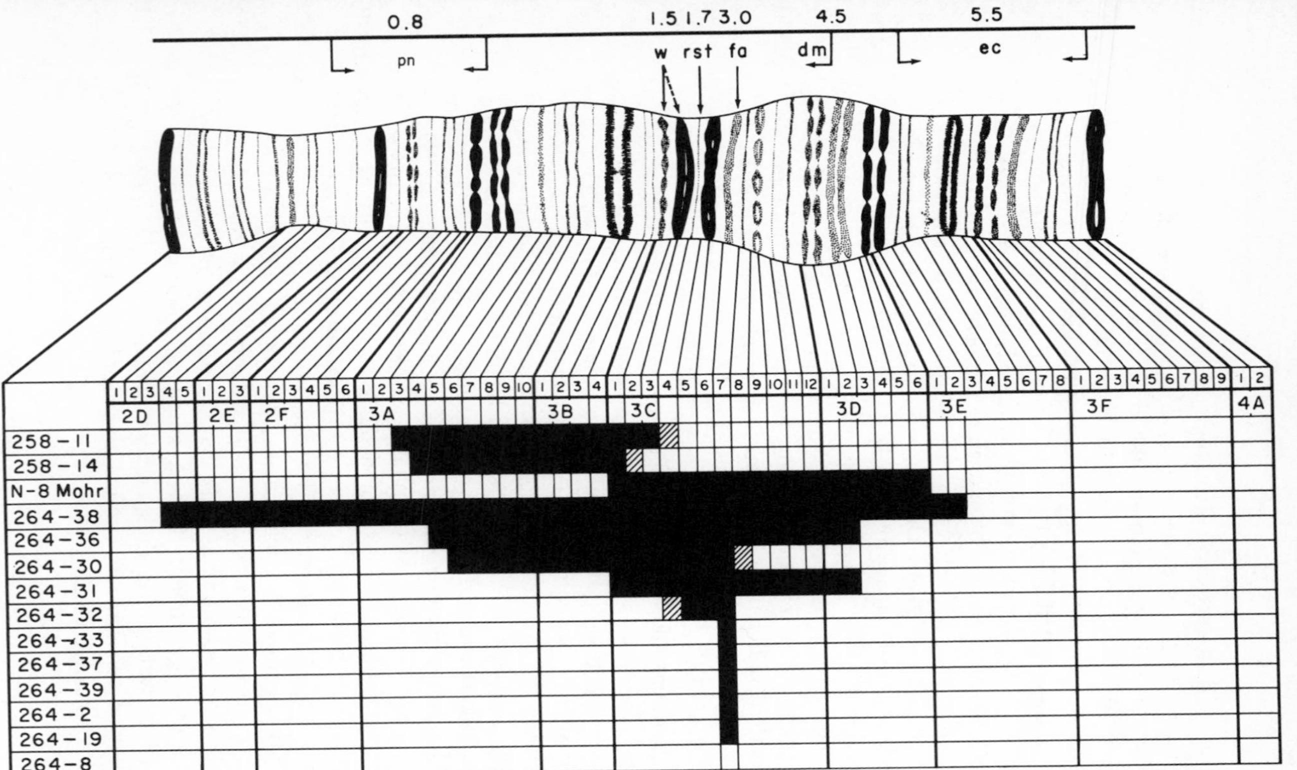

Fig. 4–14. Genetic map, corresponding segment of the salivary-gland X chromosome, and 14 deficiencies responsible for the *white-Notch* phenotype in *Drosophila melanogaster*. The black segments below indicate the extent of each deficiency and the hatched segments, possible missing bands. (*Slizynska, 1938.*)

McClintock (1938*b*) used a deficient rod and a compensating ring chromosome in maize to demonstrate that a homozygous deficiency in somatic cells was responsible for a phenotype caused by a recessive mutation in the missing chromosomal segment. The recessive mutation might have been responsible for the absence or malfunctioning of an enzyme, so that a homozygous deficiency for the locus would have produced the same phenotype. Some of the chromosomes from the breakage of the dicentric bridge at anaphase I in plants heterozygous for the reconstructed chromosome 9 had a terminal deficiency. McClintock (1944) obtained two different terminally deficient chromosomes from such plants. Seven chromosomes (Df-1) were deficient for the segment between the knob and the terminal chromomere and six chromosomes (Df-2) for both this segment and a portion of the terminal chromomere. The transmission of the Df-1 or Df-2 chromosome by the male and female gametophytes indicated that the missing segments were not essential for normal gametophyte development or function.

Plants heterozygous for different combinations of a standard chromosome 9 with the dominant or recessive allele for yellow-green seedlings (*yg-2*) and a terminally deficient chromosome (Df-1 or Df-2) or of terminally deficient chromosomes (Df-1 and Df-2) were either crossed or self-pollinated. The heterozygous or homozygous progeny were scored for phenotype and examined cytologically (Fig. 4-15). Pale yellow (*pyd*) seedlings were homozygous for a deficient chromosome 9 with a missing terminal segment (Df-1 Df-1) or had this chromosome and the other deficient chromosome with a longer deficiency (Df-1 Df-2). White seedlings (*wd*) were homozygous for the deficient chromosome with the longer deficiency (Df-2 Df-2). Finally, combinations involving the standard chromosome 9 with the dominant or recessive allele for yellow-green (*yg*) seedling and either deficient chromosome were compared in terms of phenotype and chromosomal constitution.

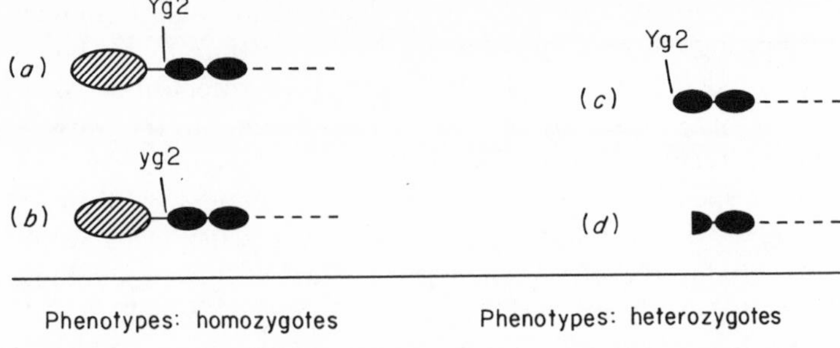

Phenotypes: homozygotes		Phenotypes: heterozygotes	
a+*a*	green seedling	*a*+*b*, *c*, or *d*	green seedling
b+*b*	yellow-green seedling	*b*+*c*	green seedling
c+*c*	pale-yellow seedling	*b*+*d*	yellow-green seedling
d+*d*	white seedling	*c*+*d*	pale-yellow seedling

Fig. 4–15. Diagrammatic representation of the short arm of chromosome 9 in maize plants with standard or terminally deficient chromosomes. (*a*) Standard chromosome with a terminal knob, two adjacent chromomeres, and the dominant allele *Yg-2;* (*b*) standard chromosome with the recessive allele *yg-2;* (*c*) terminally deficient chromosome (Df-1) lacking the knob and adjacent segment and with the dominant allele; (*d*) terminally deficient chromosome (Df-2) lacking the knob, adjacent segment, a portion of the first chromosome, and the locus. The phenotypes of seedlings resulting from homozygous or heterozygous combinations of chromosomes (*a*), (*b*), (*c*), or (*d*) are summarized in the table. (*After McClintock, 1944.*)

The pale yellow (*pyd*/*wd*) and yellow-green (*yg*/*wd*) seedlings indicated that *pyd* and *yg* were alleles of *wd*. The normal green (*yg*/*pyd*) seedling, however, indicated that *pyd* and *yg* were not allelic to each other. Two series of alleles would have had to be proposed to account for the genetic observations in terms of dominance: (1) green → *pyd* → *wd* and (2) green → *yg-2* → *wd*. The phenotype for *yg*/*pyd* could have been explained by invoking pseudoallelism. Once it was known that the *wd* and *pyd* phenotypes were expressions of different homozygous deficiencies and not of mutant genes, there was no need to construct speculative or unorthodox models to explain the genetic observations.

DUPLICATION

Several different methods are available for adding chromosomal material to a diploid species. The addition of material already present in the chromosomes constitutes a duplication. For example, an extra standard chromosome is a duplication for an entire chromosome. The addition of a specific chromosomal segment already present in the X chromosome of *D. melanogaster* will be discussed because the analysis of this duplication had a significant impact on genetic theory.

The Bar eye mutation appeared spontaneously in a wild-type stock of *D. melanogaster* and was responsible for a drastic reduction in the number of eye facets. The mutation was dominant, sex-linked, and readily assigned to a locus in the linkage map. In the course of maintaining homozygous stocks, individuals with the wild-type eye or an even greater reduced eye (double Bar) were found. In relatively large numbers of progeny the new phenotypes occurred with approximately equal frequencies (1.6×10^{-3}). While these observations suggested that Bar locus was relatively mutable, the appearance of only two mutant phenotypes in approximately equal frequencies was not easily explained. Moreover, mutant phenotypes were found in progeny from Bar females but not from Bar males. This observation suggested that mutation was not responsible for wild-type and double-Bar mutant progeny and that this difference between the sexes might be related to an unusual characteristic of *D. melanogaster;* namely, crossing-over occurs in the female but not in the male.

Sturtevant (1925) demonstrated crossing-over within the Bar locus, using females homozygous for this mutation and heterozygous for closely linked, coupled, flanking genic markers (Fig. 4-16*a*). The progeny with the wild-type or the double-Bar phenotype had one or the other flanking marker and were recombinant. To account for these observations, unequal crossing-over presumably occurred within the Bar locus, so that the reciprocal products were a recombinant chromatid with the wild-type factor and a complementary

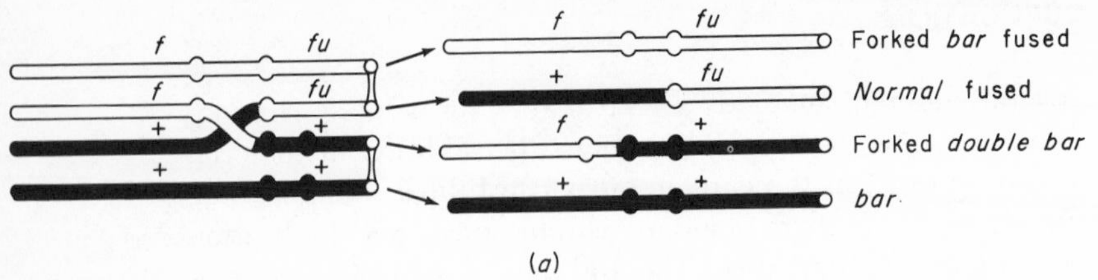

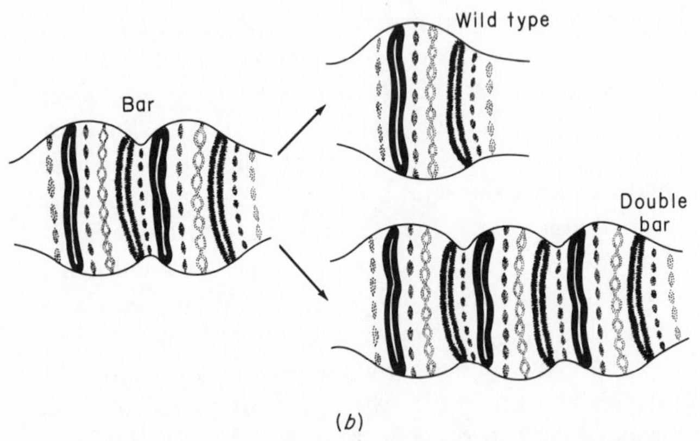

Fig. 4–16. (*a*) Diagrams illustrating the origin of normal and double-Bar progeny by crossing over in females of *Drosophila melanogaster* homozygous for Bar and heterozygous for the closely linked genes forked and fused. (*b*) Diagrams illustrating the number of 16A segments in the Bar, normal, and double-Bar salivary-gland chromosomes. (*White, 1954.*)

recombinant chromatid with the double-Bar factor. These unusual assumptions concerning crossing-over within a locus were later validated by evidence from salivary-gland chromosome analysis (Bridges, 1936).

Females heterozygous for Bar have one X chromosome

with two 16A segments in tandem and one X chromosome with only one 16A segment (Fig. 4-16*b*). This segment near the right end of the chromosome has three single bands and two doublet bands. By comparing the sequence of bands, the tandem duplication was shown to have the same sequence as the resident segment in the standard chromosome. By unequal crossing-over in homozygous Bar females, one 16A segment is transferred to a chromatid with the duplication, and eventually a chromosome with only the resident segment (wild type) and one with three segments (double Bar) are recovered. Prior to this cytological explanation for the unusual behavior of the Bar mutation, the assumption of intragenic crossing-over seemed to violate one of the contemporary dogmas of genetics.

The mean number of eye facets in flies with different numbers of 16A segments was determined, and the distribution of these segments in the two chromosomes was also noted (Fig. 4-17). For example, a fly with two 16A segments on each chromosome had 68 facets, and a fly with one 16A segment on one chromosome and three 16A segments on the other chromosome had 45 facets (Sturtevant, 1925). These observations indicated that a phenotype can be determined not only by the number and type of alleles but also by their physical location in the chromosomes. The cis-trans test of the microbial geneticist is a lineal descendant of the position-effect phenomenon first discovered for the Bar mutation.

Duplicate (15:1) and triplicate (63:1) genic ratios have

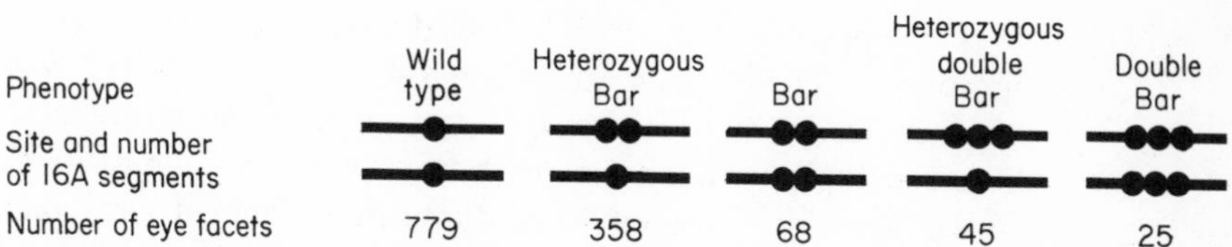

Fig. 4–17. The mean number of facets in the compound eyes of females of *Drosophila melanogaster* with different numbers of 16A segments responsible for the Bar or double-Bar phenotypes.

been reported for a few presumably diploid species. Monoploid plants can exhibit one or more bivalents, indicating that apparently nonhomologous chromosomes may have homologous segments. These genetic and cytological observations suggest that duplications may have occurred as intrachromosomal structural aberrations and had become homozygous during the evolution of diploid species. From this point of view, the standard chromosomes of contemporary species might have homozygous intrachromosomal structural aberrations, such as duplications, whose presence cannot be detected without meaningful clues from cytological or genetic observations not conforming to orthodox interpretations. Duplications provide one means for a diploid species to increase its chromosomal content without altering its chromosome number.

Duplicate genes have been reported for a number of mutant phenotypes in maize. Only individuals homozygous for the recessive alleles at different loci exhibit the mutant phenotype. Although maize is usually considered to be a diploid species with ten pairs of chromosomes, one school of thought maintains that maize might be an ancient tetraploid species derived from an interspecific hybrid of two related species with five pairs of chromosomes. Rhoades (1951) has presented an interesting discussion of duplicate genes in this species. While the loci for the duplicate genes *an-1* and *an-2* are in chromosome 9, the loci for the duplicate genes *w-5*, *w-6* and *pg-11*, *pg-12* are in different chromosomes, either 6 or 9. Breeding data indicated that *an-1* and *pg-12* are closely linked in the long arm of chromosome 9 and either *w-5* or *w-6* and *pg-11* are also closely linked in the long arm of chromosome 6. These observations suggest that a duplicated segment including the loci for *an-1* and *pg-12* is in chromosome 9 and a duplicated segment with the loci for *w-5* or *w-6* and *pg-11* is in chromosome 6. A duplicated segment presumably is responsible for the duplicate ratios, and the genes included in the segment are linked as they were in the original segment.

REFERENCES

BENZER, S. 1961. On the topography of the genetic fine structure. Proc. Nat. Acad. Sci. 47:403–415.

BRIDGES, C. B. 1936. The Bar "gene" a duplication. Science 83:210–211.

DOBZHANSKY, T. 1951. Genetics and the origin of species, 3d ed. Columbia University Press, New York.

KHUSH, G. S., AND C. M. RICK. 1967. Studies on the linkage map of chromosome 4 of the tomato and on the transmission of induced deficiencies. Genetica 38:74–94.

——— AND ———. 1968. Cytogenetic analysis of the tomato genome by means of induced deficiencies. Chromosoma 23:452–484.

MCCLINTOCK, B. 1938*a*. The fusion of broken ends of sister half-chromatids following chromatid breakage at meiotic anaphases. Missouri Agr. Exp. Sta. Res. Bull. 290.

———. 1938*b*. The production of homozygous deficient tissues with mutant characteristics by means of the aberrant mitotic behavior of ring-shaped chromosomes. Genetics 23:315–376.

———. 1941. The stability of broken ends of chromosomes in *Zea mays*. Genetics 26:234–282.

———. 1944. The relation of homozygous deficiencies to mutations and allelic series in maize. Genetics 29:478–502.

MULLER, H. J. 1928. The production of mutations by X-rays. Proc. Nat. Acad. Sci. 14:714–726.

RHOADES, M. M. 1951. Duplicate genes in maize. Amer. Natur. 85:105–110.

———. 1955. The cytogenetics of maize, pp. 123–219. *In* G. F. Sprague (ed.), Corn and corn improvement. Academic, New York.

SLIZYNSKA, H. 1938. Salivary chromosome analysis of the white-facet region in *Drosophila*. Genetics 23:291–299.

STERN, C. 1960. Principles of human genetics, 2d ed. W. H. Freeman, San Francisco.

STURTEVANT, A. H. 1925. The effects of unequal crossing over at the Bar locus in *Drosophila*. Genetics 10:117–147.

——— AND T. DOBZHANSKY. 1936. Inversions in the third chromosomes of wild races of *Drosophila pseudoobscura* and their use in the study of the history of the species. Proc. Nat. Acad. Sci. 22:448–450.

White, M. J. D. 1954. Animal cytology and evolution, 2d ed. Cambridge University Press, Cambridge.

SUPPLEMENTARY REFERENCES

Beadle, G. W., and A. H. Sturtevant. 1935. X chromosome inversions and meiosis in *Drosophila melanogaster*. Proc. Nat. Acad. Sci. 21:384–390.

Bridges, C. B. 1937. Correspondences between linkage maps and salivary chromosome structure, as illustrated in the tip of chromosome 2R of *Drosophila melanogaster*. Cytologia, Fujii jubilee volume, pp. 745–755.

Rhoades, M. M., and E. Dempsey. 1953. Cytogenetic studies of deficient-duplicate chromosomes derived from inversion heterozygotes in maize. Amer. J. Bot. 40:405–424.

Sturtevant, A. H. 1926. A crossover reducer in *Drosophila melanogaster* due to inversion of a section of the third chromosome. Biol. Zentralbl. 46:697–702.

Sutton, E. 1943. Bar eye in *Drosophila melanogaster:* a cytological analysis of some mutations and reverse mutations. Genetics 28:97–107.

5 INTERCHROMOSOMAL STRUCTURAL ABERRATIONS

A single break in two chromosomes yields four segments, each with a raw end. The appropriate fusion of these segments can produce two chromosomes, each with only one centromere and with segments from different chromosomes, that is, a reciprocal translocation of terminal segments. Although the mutual exchange of terminal segments can involve homologous chromosomes, a reciprocal translocation usually occurs between nonhomologous chromosomes. Ionizing radiation is frequently used to break chromosomes to produce reciprocal translocations. On rare occasions, two breaks in each of two chromosomes occur, and a reciprocal translocation of intercalary segments can be obtained. The commonly encountered interchromosomal structural aberration, therefore, involves two nonstandard or translocation chromosomes, resulting from a reciprocal translocation of terminal segments of nonhomologous chromosomes.

CYTOLOGICAL OBSERVATIONS

The synapsis of homologous segments of the chromosomes in a heterozygous reciprocal translocation produces a pachytene cross configuration (Fig. 5-1). In species with good pachytene chromosomes, the site of break in each nonhomologous chromosome, the length of the exchanged segments, and the chromosomes involved in the reciprocal translocation are readily determined. The salivary-gland chromosomes of species of

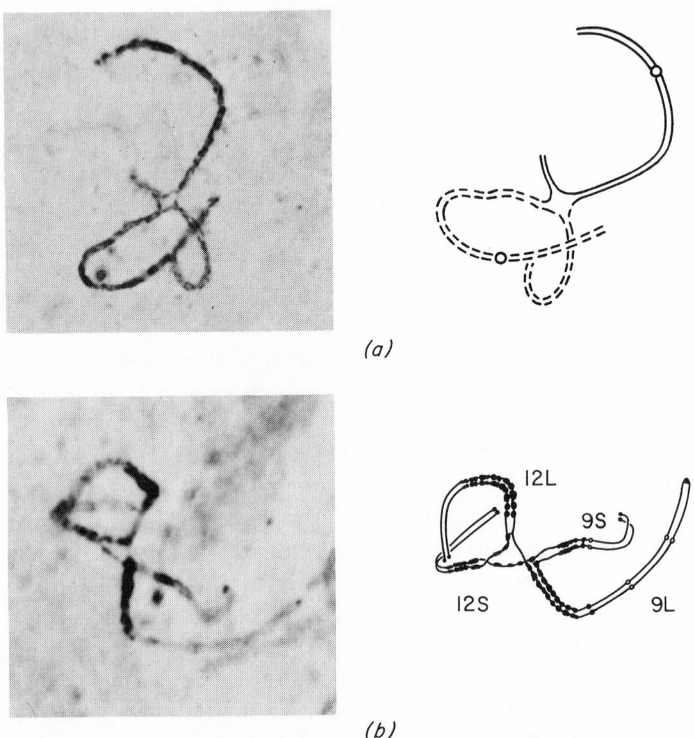

Fig. 5–1. (*a*) Pachytene cross configuration for a heterozygous reciprocal translocation involving chromosomes 8 and 10 in maize. The intersection of the cross represents the sites of break in each chromosome. (*Photograph courtesy of Dr. M. M. Rhoades; Rhoades, 1955.*) (*b*) Pachytene cross configuration for a heterozygous reciprocal translocation involving chromosomes 9 and 12 in tomato. According to the unpaired strands in the proximal region, the interchanged chromosomes are 9S•12S and 9L•12L. (*Photograph courtesy of Dr. C. M. Rick; Khush and Rick, 1967.*)

dipterans appear as arms emerging from a heavily staining body, the *chromocenter*, which includes the heterochromatic regions. In a heterozygous reciprocal translocation, the sites of break in the nonhomologous chromosomes must occur within the euchromatic regions to display the cross configuration.

The usual procedure for detecting a heterozygous reciprocal translocation cytologically requires meiocytes at diakinesis or metaphase I. Four chromosomes generally form either a ring or chain of four chromosomes, known as an *interchange complex*. The formation of chiasmata in the homologous segments of the cross configuration yields an association of four chromosomes at metaphase I. When each arm of the cross has at least one chiasma, the complex appears as a ring, and if one arm lacks a chiasma, the complex occurs as a chain. In some cases, the terminalization of chiasmata can be responsible for a chain of three chromosomes and a univalent or two pairs of chromosomes.

The frequencies of pollen mother cells with a ring or chain interchange complex or two pairs of chromosomes in lieu of the complex are determined by the relative lengths of the translocated segments in maize. When most of the cells have a ring complex, both segments are relatively long; when the cells have a ring or chain complex, one segment is relatively long and the other relatively short; when the cells have mostly the chain complex, one segment is short; and when the cells have either the chain complex or the two pairs of chromosomes, one segment is relatively short and the other segment very short. These relations may be useful in determining the relative length of the translocated segments of interchange complexes in other species.

The orientation of the ring or chain interchange complex at metaphase I is pertinent to an understanding of both the fertility of translocaton heterozygotes and the transmission of genes in the chromosomes of the complex. Although the orientation of the complex will be discussed in terms of a ring of four chromosomes, the chain and the ring behave in the same manner. The complex occurs either as an open ring or a zigzag (infinity-sign) ring in which two centromeres go to one pole and the other two centromeres go to the opposite pole. The open ring is an adjacent orientation (Fig. 5-2*a* and *c*) and the zigzag ring an alternate orientation (Fig. 5-2*b*) of the centromeres going to the same pole. Furthermore, the adjacent orientation may separate homologous centromeres (adjacent 1) or nonhomologous centromeres (adjacent 2) at anaphase I. When the centromeres of the chromosomes are not coordinated as in an open-ring configuration at metaphase I (Fig. 5-2*c*), the chromosomes can be distributed in a 3:1 rather than the usual 2:2 manner at anaphase I.

The fertility of translocation heterozygotes is related to the frequency of meiocytes with the alternate orientation of the interchange complex at metaphase I. In plant species, the translocation heterozygote can be semisterile with 50 percent aborted pollen grains and ovules or almost as fertile as standard plants. In the semisterile plants, approximately 50 percent of the meiocytes have an alternate orientation, and in the second category, significantly more than 50 percent of the meiocytes have the alternate orientation. The orientation of the interchange complex at metaphase I is said to be *directed* when significantly more than 50 percent of the meiocytes have

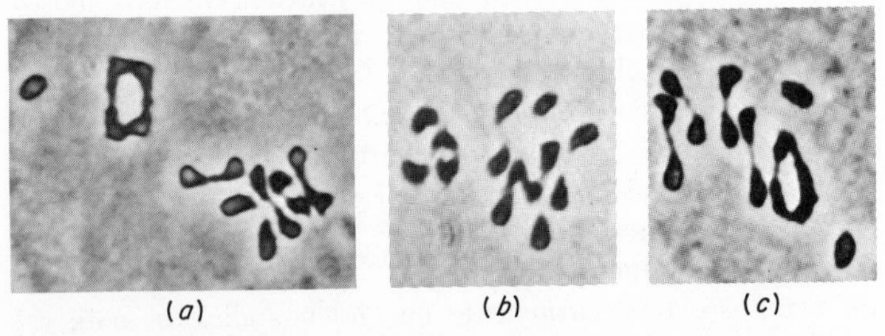

Fig. 5–2. Heterozygous reciprocal translocations at metaphase I in *Collinsia tinctoria*, $n = 7$. (*a*) One interchange complex displaying the adjacent orientation in a plant with an extra chromosome ($2n + 1$). (*b*) A ring and a chain interchange complex exhibiting the alternate orientation. (*c*) One ring complex with a discordant (3:1) adjacent orientation. (*Photographs courtesy of Dr. K. S. Rai.*)

the alternate orientation. A species has either a directed or a nondirected orientation of the complex but, as a rule, not both.

The distribution of the chromosomes from the interchange complex at anaphase I determines the chromosomal constitution of the gametophytes or gametes (Fig. 5-3). Meiocytes with the adjacent orientation at metaphase I yield four deficient haploid nuclei, and those with the alternate orientation yield two haploid nuclei with both standard chromosomes and two haploid nuclei with both translocation chromosomes. The deficient gametophytes usually abort. Although the deficient gametes of animals are functional, the deficient zygotes generally do not yield viable embryos or individuals. When gametes with different deficiencies are involved in fertilization, the zygote develops normally. Consequently, the percentage of aborted pollen grains and ovules is directly related to the frequency of meiocytes with the adjacent orientation of the interchange complex at metaphase I.

A self-pollinated semisterile plant produces approximately equal numbers of semisterile or fertile progeny; the progeny from a cross between a semisterile and a standard plant include approximately equal numbers of semisterile and fertile individuals. These breeding data are readily interpreted in terms of the chromosomal constitution of the gametophytes and zygotes (Fig. 5-3). The fertile progeny from a self-pollinated semisterile plant are either standard individuals or translocation homozygotes in equal numbers. The translocation homozygotes display only bivalents but can be distinguished from the standard plants by breeding tests or sometimes by chromosome morphology. When the fertile F_2 plants are crossed with standard plants, all the progeny are either fertile or semisterile, indicating that the fertile parents had standard chromosomes or homozygous translocation chromosomes, respectively. If the reciprocal translocation grossly alters the morphology of the standard chromosomes, the translocation homozygote will have two bivalents not found in standard plants. The procedures used for semisterile plants cannot be applied directly to animals with a heterozygous reciprocal

Pachytene

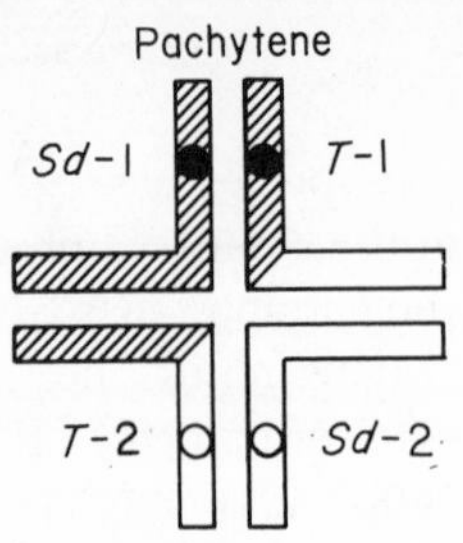

Metaphase I orientation

Alternate

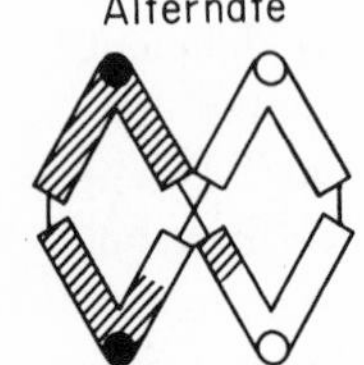

Adjacent 1

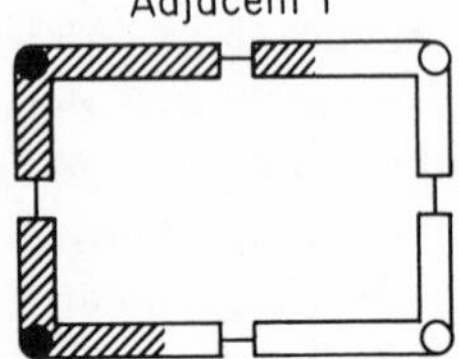

Adjacent 2

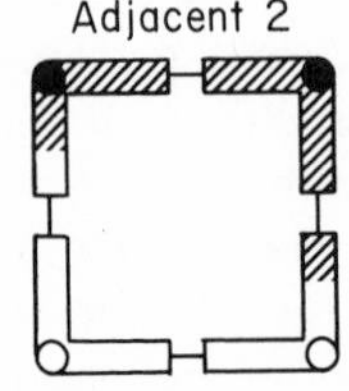

Anaphase I distribution

Sd-1 *Sd*-2 *T*-1 *T*-2

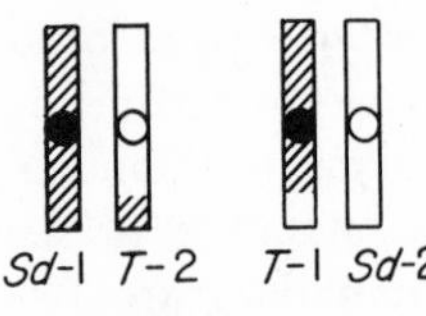

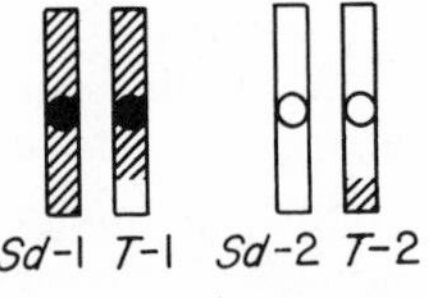

Gametophytes

Viable | Aborted | Aborted

Self-pollination	Transmission *Sd*-1 *Sd*-2	*T*-1 *T*-2
Sd-1 *Sd*-2	*Sd*-1 *Sd*-1 *Sd*-2 *Sd*-2 bivalents, fertile	*Sd*-1 *T*-1 *Sd*-2 *T*-2 interchange complex semisterile
T-1 *T*-2	*Sd*-1 *T*-1 *Sd*-2 *T*-2 interchange complex semisterile	*T*-1 *T*-1 *T*-2 *T*-2 bivalents, fertile
Test cross *Sd*-1 *Sd*-2	*Sd*-1 *Sd*-1 *Sd*-2 *Sd*-2 bivalents, fertile	*Sd*-1 *T*-1 *Sd*-2 *T*-2 interchange complex semisterile

translocation. The functional deficient gametes preclude a direct examination, and the progeny of appropriate crosses must be used to determine the transmission of the chromosomes of the interchange complex.

While the semisterile plants have either the alternate or adjacent orientation of the interchange complex at metaphase I, two types of adjacent orientation can occur. Occasionally, a reciprocal translocation produces an interchange complex in which each chromosome is obviously different in morphology. In such cases, the two types of adjacent orientation can be identified by direct observation. The two types can also be identified by another and more general method by relating the disjunction of homologous centromeres in a standard and a translocation chromosome to the adjacent-1 orientation and the nondisjunction of homologous centromeres to the adjacent-2 orientation.

In a diploid plant species with one pair of nucleolus-organizing chromosomes, each of the four microspores in a quartet has a single nucleolus in the nucleus. An interchange complex involving this chromosome yields microspores with diffuse nucleolar material or with two nucleoli (Fig. 5-4). Assuming no crossing-over in the intercalary segment between the centromere and site of break in the translocation chromosome, the alternate and adjacent-1 orientations of the complex at metaphase I produce quartets with one nucleolus in each nucleus, and the adjacent-2 orientation produces quartets in which two nuclei have diffuse nucleolar material and two nuclei have one large nucleolus or two nucleoli. Detecting the adjacent-2 orientation and determining its frequency require the frequencies of meiocytes with the alternate or adjacent orientation and quartets with diffuse nucleolar material in the microspore nuclei.

Fig. 5–3. The relation between the orientation of an interchange complex of four chromosomes responsible for semisterility at metaphase I and the transmission of a heterozygous reciprocal translocation to progeny after self-pollination or a testcross.

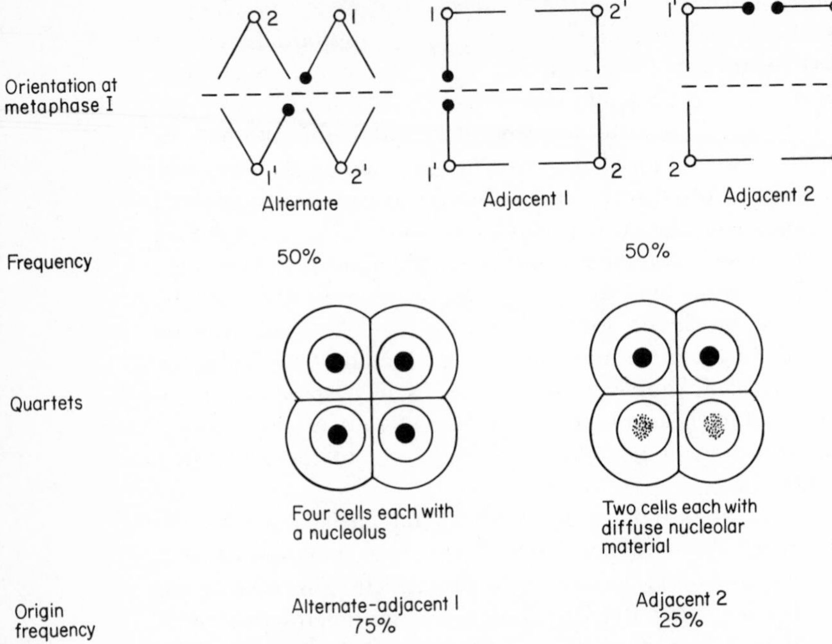

Fig. 5–4. Scheme illustrating the origin of quartets at the end of microsporogenesis with one nucleolus in each nucleus or with diffuse nucleolar material in two adjacent nuclei in a translocation heterozygote with an interchange complex involving a chromosome with a terminal nucleolus-organizing region. (*Garber, 1948.*)

A crossing-over in the intercalary segments of an interchange complex can yield different types of quartets with respect to the number of viable microspores as well as the number and relative position of microspores with diffuse nucleolar material (Fig. 5-5). Burnham (1949) used an interchange complex involving chromosomes 5 and 6 in maize to detect the products of crossing-over in the intercalary segments. The quartets were scored for the number of nucleoli, the presence of diffuse nucleolar material, and the relative position of the different types of nuclei. The data indicated that a crossing-over in the intercalary segment was followed by the disjunction of the homologous centromeres (adjacent 1) so that the adjacent-2 orientation was not found at metaphase I. While the number of viable or inviable gametophytes could be predicted, they could not be identified at the quartet stage in maize. Kihara and Shimotsuma (1967) obtained such data for interchange heterozygotes in the watermelon, where the aborted gametophytes are retained in a quartet, and these data supported the predicted observations. Zimmering (1955) used genetic data to demonstrate that crossing-over in the intercalary segment of a heterozygous reciprocal translocation in *Drosophila melanogaster* resulted in the separation of homologous centromeres (adjacent 1) at anaphase I.

GENETIC OBSERVATIONS

The semisterility associated with an interchange complex in maize constitutes a phenotypic alteration to contrast with the fertility of plants with standard chromosomes. The progeny from a cross between a translocation heterozygote and a standard individual include approximately equal numbers of semisterile and fertile individuals, thereby simulating a monohybrid testcross. Consequently, a translocation heterozygote also heterozygous for a genic marker can be used in a testcross to determine whether the gene is in one of the chromosomes of the interchange complex (Fig. 5-6). When the gene is not in either chromosome of the complex, the testcross progeny include four classes of offspring in approximately equal numbers: (1) semisterile, dominant phenotype, (2) semisterile, recessive phenotype, (3) fertile dominant phenotype, and (4) fertile recessive phenotype. When the locus is in one of the chromosomes of the interchange complex, the frequencies of the four classes are determined by the frequency of crossing-over in the intercalary segment between the locus and the site of break in the translocation chromosome. It should be em-

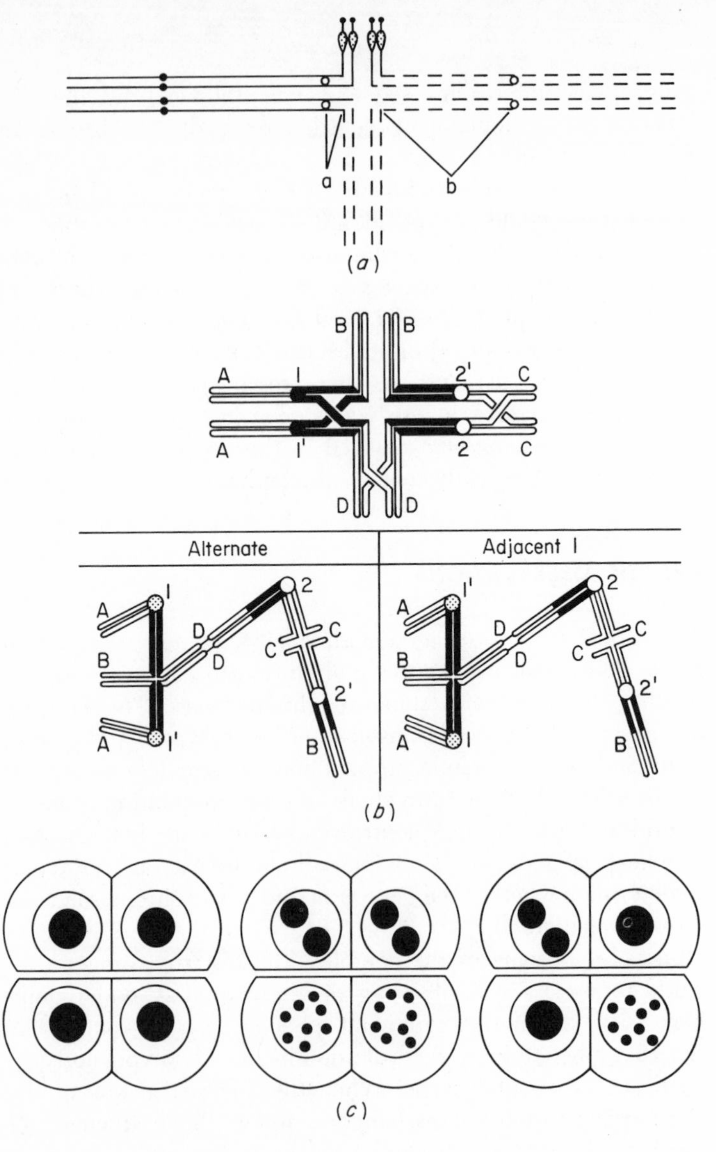
a
b
(a)
B
B
A
1
2'
C
A
1'
2
C
D
D
Alternate
Adjacent 1
A
1
2
D
C
C
B
D
2'
A
1'
B
A
1'
2
D
C
C
B
D
2'
A
1
B
(b)
(c)

phasized that a linkage between the gene and semisterility does not indicate which chromosome in the complex has the locus.

The inheritance of two genic markers known to be in the nonhomologous chromosomes of an interchange complex is determined by the frequency of crossing-over in the respective intercalary segments, thereby simulating a linkage for the two markers. This phenomenon has been termed *pseudolinkage*. Although recombinant nuclei are produced from the adjacent orientation of the interchange complex at metaphase I, the gametophytes with the deficient nuclei abort. The recombinant progeny represent the products of crossing-over in the inter-

---

Fig. 5–5. (*a*) Pachytene configuration of a translocation heterozygote involving chromosomes 5 and 6 of maize. The break in chromosome 5 is in the long arm, and the break in chromosome 6 in the short arm and near the centromere. The intercalary segment *a* between the centromere 6 and the translocation point is very short, so that the crossing-over in this segment occurs infrequently; the other intercalary segment (*b*) between the centromere and the translocation point in chromosome 5 is sufficiently long to permit frequent crossing-over. (*Burnham, 1949.*) (*b*) Crossing-over in the intercalary segment (*solid line*) between one centromere and interchange point and in the segment distal from the other centromere (pachytene). The alternate and adjacent-1 orientations at metaphase I yield the same configuration, but different chromosomes proceed to the same pole at anaphase I. (*Lewis and John, 1963.*) (*c*) Principal types of microspore quartets from plants heterozygous for reciprocal translocations involving chromosome 6 in maize. *Left quartet:* Each spore has one nucleolus-organizing region, and therefore one nucleolus comes from the alternate and adjacent-2 segregation when the break for chromosome 6 is in the short arm. When the interchange point is in the long arm, both alternate and adjacent-1 segregations yield this type of quartet. *Center quartet:* Two spores, each with two nucleoli, have two nucleolus-organizing regions, and two spores, each with diffuse nucleolar material, lack a nucleolus-organizing region. This type of quartet comes from an adjacent-1 segregation when the break in chromosome 6 is in the short arm and from an adjacent-2 segregation when the break is in the long arm. *Right quartet:* One spore with two nucleoli, two each with one nucleolus, and one with diffuse nucleolar material occur in translocation heterozygotes with the break in the short arm of chromosome 6 when there is a crossover in the intercalary region followed by either an alternate or adjacent-1 segregation. (*Rhoades, 1955; after Burnham, 1949.*)

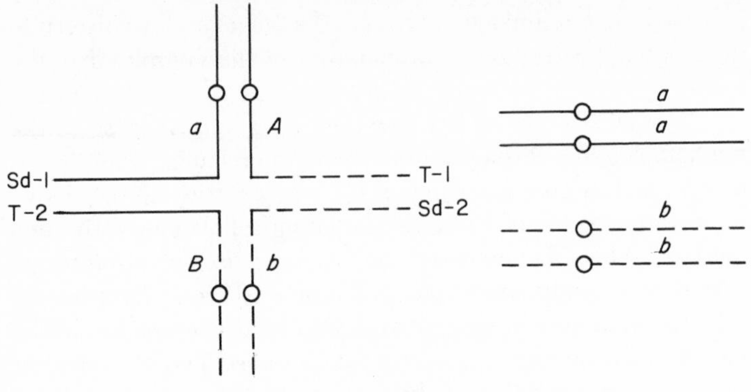

Relative Frequencies

	Semisterile				Fertile			
	AB	*Ab*	*aB*	*ab*	*AB*	*Ab*	*aB*	*ab*
Standard heterozygote								
	—	—	—	—	1	1	1	1
Translocation heterozygote								
No crossing over *A–a* or *B–b*	1	0	0	0	0	0	0	1
Crossing over								
A–a	High	0	Low	0	0	Low	0	High
B–b	High	Low	0	0	0	0	Low	High
A–a and *B–b*	High	Low	Low	Low	Low	Low	Low	High

Fig. 5–6. Relative frequencies of semisterile or fertile progeny from a testcross between a standard individual homozygous for genes in nonhomologous chromosomes and an individual heterozygous for a reciprocal translocation involving these chromosomes and for the pertinent genes. Crossing-over may or may not occur in the segment between the locus for each gene and the site of the interchange in each chromosome.

calary segments in meiocytes with the alternate orientation or, in certain cases, an adjacent-1 orientation of the interchange complex.

Reciprocal translocations have been useful cytogenetic

tools for solving a number of problems. For example, linkage maps can be assigned to specific chromosomes by detecting linkages with interchange complexes known to involve particular chromosomes. In such species as maize or tomato, the linkage maps can also be oriented in a standard sequence directly related to the corresponding chromosomes. A direct examination of the pachytene cross configuration permits the sites of break in the translocation chromosomes to be assigned to the short or long arm of the chromosome and proximal or distal to the centromere. Rhoades (1933) used a reciprocal translocation in maize with one break in the short arm of chromosome 5 and the other break in the long arm of chromosome 2; both breaks were close to the respective centromere. The heterozygous reciprocal translocation carrying a number of linked genes in each chromosome was crossed to a standard individual homozygous for the recessive alleles, and the progeny were scored for dominant or recessive semisterile or fertile plants to determine the linkage values for each locus and the site of break in the translocation chromosomes. The recombination values indicated which genes were close to the centromere in each chromosome and which genes were on the short or long arm of each chromosome. Anderson and Randolph (1945) summarized the available data from reciprocal translocations and recombination values to locate the centromeres in the linkage maps of maize.

The linkage maps for *D. melanogaster* were directly related to the corresponding chromosomes in terms of the sites of the loci in each group and chromosome. The relative distances between the loci in the maps did not always agree with those for the sites in the chromosomes, depending on the region of the chromosome. The discrepancies were greatest at the center of the two large autosomes and at one end of the X chromosome, and these regions were the sites of the centromeres in each case; one crossing-over unit was equivalent to a relatively greater physical distance in these regions than in the other regions of the chromosomes. Consequently the frequency of crossing-over in a chromosomal region is related to its proximity to the centromere, at least in *D. melanogaster.* Other morphological characteristics of chromosomes can be expected to influence the frequency of crossing-over in the immediate neighborhood of the characteristic.

CROSSING-OVER AND CHROMOSOMAL EXCHANGE

While it had seemed reasonable to equate the crossing-over of linked genes with a chromosomal exchange, the supporting experimental evidence was not available until 1931. Creighton and McClintock (1931) and Stern (1931) used reciprocal translocations in maize and *D. melanogaster,* respectively, to provide the visual evidence for chromosomal exchange. The experimental design required two linked genes which were reasonably close to minimize double crossing-over between them and a visual chromosomal marker at each end of the segment of the linkage map with the genic markers. In maize, the chromosomal markers were the presence or absence of a large knob at the end of the short arm of chromosome 9 and a reciprocal translocation involving this chromosome and chromosome 8; the genic markers were colorless aleurone (*c*) and waxy endosperm (*wx*). The heterozygous reciprocal translocation was crossed with a knobless standard parent with the *c Wx/c wx* genotype, rather than the recessive homozygote. Although relatively few genetic recombinants were obtained, they had the appropriate combination of chromosomal markers to demonstrate the correlation between genetic crossing-over and chromosomal exchange.

The small mitotic chromosomes of *D. melanogaster* required an ingenious manipulation of reciprocal translocations to yield the necessary cytological markers for detecting the products of chromosome exchange (Fig. 5-7). Stern obtained a translocation X chromosome with an easily detected segment of the Y chromosome and a reciprocal translocation between the X and chromosome IV, so that a female with this reciprocal translocation had nine rather than eight chromosomes. The

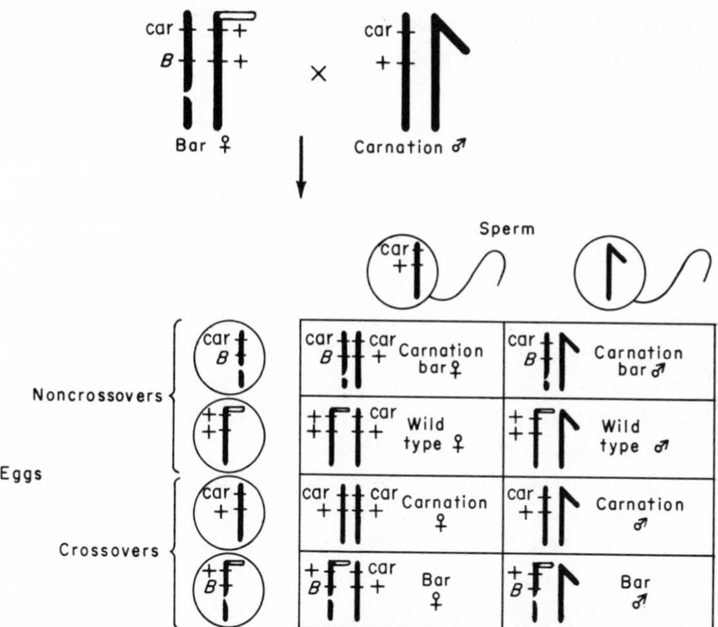

Fig. 5–7. Correlation between recombination of sex-linked genes and chromosomal exchange in *Drosophila melanogaster*. One X chromosome had a segment (*outlined*) of the Y chromosome, and the other X chromosome was involved in a reciprocal translocation with chromosome IV. Progeny with eight or nine chromosomes and with a standard or the X-Y can be readily distinguished by a cytological examination of individuals with different phenotypes. (*After Stern, 1931.*)

gene markers were placed in the segment of the X chromosome relatively close to the Y segment in this translocation chromosome. The progeny from the testcross included the parental genotypes with eight or nine chromosomes and recombinants also with eight or nine chromosomes. One recombinant, however, had eight chromosomes, including both standard X chromosomes or one X and Y chromosome, and the other recombinant had nine chromosomes, including a standard X or the Y chromosome and the X-Y translocation chromosome.

CYTOLOGICAL IDENTIFICATION OF CHROMOSOMES IN INTERCHANGE COMPLEXES

Many plant and animal species do not have suitable pachytene chromosomes for identifying the chromosomes in an interchange complex. In species with extensive linkage maps, pseudolinkage can be used to assign a linkage group to one of the two chromosomes in the complex. This method, however, requires a number of crosses with different heterozygous reciprocal translocations to detect pseudolinkage for the genic markers reasonably close to the breaks in the translocation chromosomes. Cytological identification of the chromosomes in interchange complexes is more efficient when a set of translocation tester strains is available.

A collection of heterozygous reciprocal translocations is assembled and examined at diakinesis to determine which interchange complexes are associated with the nucleolus. Such reciprocal translocations involve the nucleolus-organizing chromosome. The other translocation heterozygotes are crossed in all combinations, and the progenies are examined cytologically to determine which hybrids have two interchange complexes or one interchange complex of six chromosomes at diakinesis or metaphase I. When a hybrid has two interchange complexes, the chromosomes involved in the reciprocal translocation of one parent are arbitrarily assigned numbers 1 and 2 and those of the second parent, numbers 3 and 4. When a hybrid involving T1-2 or T3-4 has two interchange complexes, the chromosomes in the third reciprocal translocation are arbitrarily assigned numbers 5 and 6. Finally, a hybrid with an interchange complex of six chromosomes indicates that the reciprocal translocation in each parent involved a common chromosome. A set of tester strains is selected so that a minimum number of crosses will be needed to identify the chromosomes in almost any reciprocal translocation. Burnham, White, and Livers (1954) used a set of five tester strains in barley ($n = 7$) to identify the chromosomes in a number of reciprocal translocations (Table 5-1).

TABLE 5-1

Critical Chromosome Associations at Diakinesis or Metaphase I in Hybrids between the Tester Stocks and between Tester Stocks and Other Translocation Heterozygotes to Determine the Chromosomes Involved in the Translocation Heterozygotes for a Species with $n = 7$ Chromosomes

	Testers Indicating Translocation Chromosomes *					
	a-b	b-d	c-e	e-f	c-d	Translocation Chromosomes
Testers:						
a-b						
b-d	1⊙6					
c-e	2⊙4	2⊙4				
e-f	2⊙4	2⊙4	1⊙6			
c-d	2⊙4	1⊙6	1⊙6	2⊙4		
Unknowns:						
1	1⊙6	1⊙6	1⊙6	1⊙6		b-e
2	1⊙6	1⊙6	2⊙4	1⊙6		b-f
3	1⊙6	1⊙6	2⊙4	2⊙4	2⊙4	b-g
4	7^{II}	1⊙6	2⊙4	2⊙4		a-b

* Interchange complexes involving the nucleolus-organizing chromosome (*g*) are associated with the nucleolus at diakinesis.

Source: After Burnham, White, and Livers, 1954.

STERILITY AND RECIPROCAL TRANSLOCATIONS

Translocation heterozygotes causing semisterility yield semisterile and fertile progeny when self-pollinated or crossed to a standard plant. Consequently, the progeny can be scored by determining the percentage of aborted pollen grains without resorting to a cytological examination to detect the presence or absence of the interchange complex. Progeny with more or less than the expected 50 percent aborted pollen grains, however, are occasionally found, and such individuals may be trisomics ($2n + 1$). The distribution of the chromosomes in the interchange complex at anaphase I is occasionally 3:1

rather than the usual 2:2, thereby producing spores with an extra chromosome ($n + 1$). Such spores have one standard and both translocation chromosomes or one translocation and both standard chromosomes. The fertility of these trisomic progeny can be calculated by assuming that the extra chromosome is always included in half of the spores which are always functional.

Crosses between one translocation heterozygote and a presumably standard plant may yield progeny with 75 percent aborted pollen grains. These progeny can have either two interchange complexes of four chromosomes or one interchange complex of six chromosomes. Although such individuals are readily identified by a cytological examination, they can also be distinguished by noting the fertility of the progeny from crosses with a known standard plant (Table 5-2).

In a number of species, particularly in the Onagraceae, the interchange complexes exhibit a directed orientation at metaphase I; that is, the frequency of pollen mother cells with the alternate orientation is significantly greater than that of cells with the adjacent orientation. Plants with a directed orientation of the complex have a significantly higher percentage of viable pollen grains and ovules than plants with a nondirected orientation (semisterility). As a rule, interchange complexes in one species have either a directed or a nondirected orientation at metaphase I. One exception has been found in *Collinsia heterophylla* ($n = 7$). Reciprocal translocations were obtained in this species by colchicine treatment (Soriano, 1957) or by ionizing radiation (Garber and Dhillon, 1961) and compared by noting (1) the frequencies of pollen mother cells with a ring or chain complex or two pairs of chromosomes in lieu of the complex at metaphase I, (2) the orientation of the complexes at metaphase I, and (3) the percentage of aborted pollen grains in translocation heterozygotes. Plants with the colchicine-induced complexes had two pairs of chromosomes or a ring or chain complex with a nondirected orientation at metaphase I, and plants with radiation-induced complexes had mostly the ring or chain complex with a

TABLE 5-2

The Origin of Viable and Aborted Gametophytes from a Maize Plant (A) Heterozygous for Two Independent Interchange Complexes, Each with Four Chromosomes, and from a Plant (B) Heterozygous for an Interchange of Six Chromosomes from a Hybrid between Translocation Heterozygotes Involving a Common Chromosome

A. Two Independent Interchange Complexes

Orientation at Metaphase I			
T1-2	T3-4	Spores	%
Alternate	Alternate	Viable	25
Alternate	Adjacent	Abort	25
Adjacent	Alternate	Abort	25
Adjacent	Adjacent	Abort	25

B. Interchange Complex of Six Chromosomes

Chromosomal Constitution	Spores	%
a-d-e	Viable	12.5
b-c-f	Viable	12.5
a-d-f	Abort	12.5
a-c-e	Abort	12.5
a-c-f	Abort	12.5
b-d-e	Abort	12.5
b-d-f	Abort	12.5
b-c-e	Abort	12.5

directed orientation at this stage. Furthermore, plants with a colchicine-induced complex were semisterile, and those with a radiation-induced complex had a significantly higher fertility.

To account for the directed or nondirected orientation of

the different types of interchange complexes in *C. heterophylla*, it was necessary to consider the possible sites of breakage (Fig. 5-8). The chromosomes of this species are metacentric and have not more than one chiasma in each arm. Each chiasma was assumed to form in a relatively small segment which was either terminal or subterminal. Ionizing radiation presumably broke chromosomes in the long segment between the chiasma-forming segments. Consequently, a chiasma could be produced in each arm of the pachytene cross configuration. Colchicine presumably broke chromosomes in the chiasma-forming segments. Because one chromosome arm can have only one chiasma, the first chiasma near the sites of breakage determines the location of the second chiasma. Although the lengths of the translocated segments differ for the two types of interchange complexes, it is not clear how these differences might be related to the nondirected or directed orientation of the complexes at metaphase I.

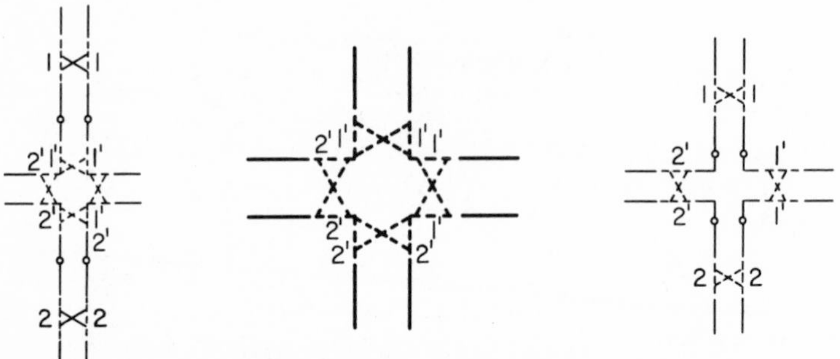

Fig. 5–8. Diagrams illustrating pachytene associations for heterozygous reciprocal translocations in *Collinsia heterophylla* produced by colchicine (*left, center*) or by ionizing radiation. Interrupted chromosomal segments represent the sites of chiasma formation; homologous segments have the same number; and the dotted crosses indicate the potential chiasmata. (*Garber and Dhillon, 1961.*)

RECIPROCAL TRANSLOCATIONS INVOLVING SUPERNUMERARY CHROMOSOMES

The standard (A) and supernumerary (B) chromosomes of maize have been involved in reciprocal translocations by means of ionizing radiation and plants with up to 20 B chromosomes (Roman, 1947). The heterozygous A-B reciprocal translocation with 25 percent pollen abortion are readily distinguished from the semisterile plants heterozygous for A-A reciprocal translocations. In an A-B translocation heterozygote, one translocation chromosome (A^B) has a B segment associated with an A centromere, and the other translocation chromosome (B^A) has an A segment with a B centromere. Spores from one adjacent orientation of the complex yielding a deficiency for the B segment do not produce aborted pollen grains.

The standard chromosome in an A-B interchange complex is readily identified at pachytene in maize. Furthermore, the linkage groups have been assigned to specific chromosomes, and many genes have been localized in the short or long arm of each chromosome as well as in particular regions of the arms. In one A-B translocation chromosome (B^4), a terminal segment of chromosome 4 with the dominant allele for the recessive endosperm phenotype sugary (*su*) was associated with a B centromere. A cytogenetic study of this B^A and other translocation chromosomes indicated that one of the two sperm nuclei of the male gametophyte was more likely to be included in the zygotic nucleus (preferential fertilization) than the other sperm nucleus.

The cross between a standard homozygous recessive seed parent (4*su* 4*su*) and the B-4 translocation heterozygote with the dominant alleles gave seeds with the following genotype and chromosomal constitution: $4su/4^B/B^4Su/B^4Su$. The nondisjunction of the B^4Su translocation chromosome at the second mitotic division in the development of the male gametophyte

1. Meiosis

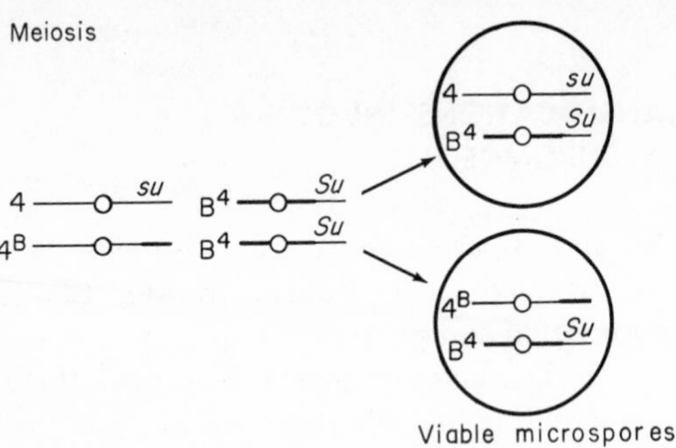

2. Nondisjunction at second mitotic division

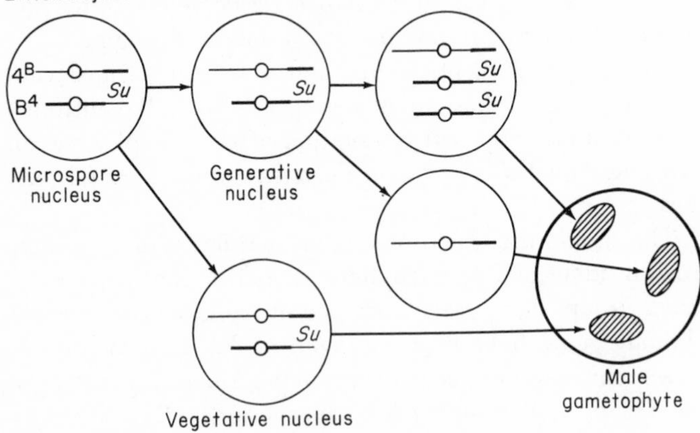

3. Double fertilization

Sperm nuclei	Polar nuclei	Egg nucleus	Seed genotypes	Seed phenotypes
B^4*Su* B^4*Su* A^B A^B	*su su*	 *su*	*Su Su su su* *su*	Nonsugary
A^B B^4*Su* B^4*Su* A^B	*su su*	 *su*	*su su* *Su Su su*	Sugary
B^4 *Su* 4^B B^4 *Su* 4^B	*su su*	 *su*	*Su su su* *Su su*	Nonsugary

produced a sperm nucleus with $4^B/B^4Su/B^4Su$ which fertilized the egg (4*su*). Consequently, the nondisjunction of B chromosomes at the second mitotic division is a property of the B centromere (and perhaps adjacent region), and not of the entire chromosome, which is expressed during the development of the male but not of the female gametophyte in maize.

Reciprocal crosses between homozygous recessive standard plants (4*su* 4*su*) and plants with $4su/4^B/B^4Su/B^4Su$ gave only nonsugary seeds when the standard plants provided the pollen and sugary seeds when the standard plants were the female parent. The sugary seeds had two B^4 chromosomes in the embryo and no B^4 chromosomes in the endosperm (Fig. 5-9). Some nonsugary seeds had a B^4 chromosome in both embryo and endosperm, indicating that the mitotic nondisjunction had not occurred in all the microspores. The mitotic nondisjunction of the B^4 chromosome rarely occurred in microspores without the corresponding 4^B translocation chromosome. This observation indicated that the B centromere was responsible for the mitotic nondisjunction of the B^4 chromosome but the rest of the heterochromatic B chromosome played a significant role. Finally, the sperm nucleus with $4^B/B^4/B^4$ fertilized the egg nucleus more frequently than it was incorporated into the primary endosperm nucleus (preferential fertilization).

CYTOGENETICS OF *Oenothera*

In 1901, de Vries proposed the mutation theory to account for the infrequent appearance of mutants in progeny from self-pollinated plants of *Oenothera lamarckiana.* Many mutants differed from the typical plants in a number of seemingly un-

---

Fig. 5–9. The inheritance of sugary seed (*su*) in maize in a cross between a standard mutant (4*su*/4*su*) female and a dominant pollen parent with A-B reciprocal translocation chromosomes $4su/4^B/B^4Su/B^4su$. (*After Roman and Ullstrop,* 1951.)

related characteristics. Although most of the mutations were transmitted when the mutants were self-pollinated, the segregating populations did not fit Mendelian ratios. Crosses between *O. lamarckiana* and related "species" as well as between "species" gave segregating populations which did not fit Mendelian models and seemed to be directly related to the "species" involved in each cross. While the genetic analysis of most plant and animal species were yielding data interpreted by simple or complex ratios or by linkage, genetic studies in *O. lamarckiana* and related "species" continued to give unorthodox results. The *Oenothera* problem was eventually solved by applying several discoveries from other species, particularly the cytogenetics of reciprocal translocations. Although specific epithets will be retained in discussing *Oenothera* cytogenetics, they have been abandoned by taxonomists and have only historical value.

Self-pollinated plants of *O. lamarckiana* gave essentially similar progeny, suggesting that most loci were homozygous. Yet crosses to other species yielded distinctly different types of progeny. These observations were explained by assuming that *O. lamarckiana* produces two genetic types of gametic nuclei responsible for different constellations of phenotypes which are kept intact in each generation. The constellation has been termed a *Renner complex* in honor of the early *Oenothera* geneticist who called attention to this phenomenon. Each complex has been assigned a name to identify a specific constellation of phenotypes. For example, *O. lamarckiana* is a true-breeding hybrid containing a *velans* and a *gaudens* complex. At the time, a true-breeding hybrid was a paradox not easily explained by orthodox genetic concepts.

Self-pollinated plants of *O. lamarckiana* set approximately half as many seeds as are found when this species is the female parent in crosses with other species. These results were explained in terms of a balanced lethal system proposed by Muller (1918) to account for a similar situation in *D. melanogaster*. According to this explanation, the *gaudens* and *velans* complexes have apparently linked recessive lethal genes in repulsion. Consequently, 50 percent of the zygotes from a self-pollinated plant of *O. lamarckiana* are homozygous for the *gaudens-gaudens* or *velans-velans* complexes and therefore homozygous for each recessive zygotic lethal, and 50 percent of the zygotes are *gaudens-velans* and heterozygous for the recessive zygotic lethals. This interpretation accounted for the different percentages of seed set for self-pollinated and outcrossed plants of *O. lamarckiana*. The concept of zygotic lethal genes was later extended to pollen or egg lethals to explain other observations from certain crosses between species (Table 5-3). Renner (1921) investigated the nature of pollen and egg lethals by classical methods.

One species (*O. hookeri*) with seven bivalents provided the experimental material for understanding the nature of egg lethality. The quartet of four megaspores is formed in the usual linear array, and the terminal megaspore at the micropylar end develops into the female gametophyte. Breeding data had already indicated that the *curvans* and not the *rigens* complex in *O. muricata* carried an egg lethal. In this species, 50 percent of the gametophytes developed from the basal megaspore and the other 50 percent from the megaspore at the opposite end. These observations were explained by assuming a developmental competition between terminal megaspores with different complexes. When the *rigens* complex is in the basal megaspore, this megaspore yields the gametophyte; when the *rigens* complex is in the other terminal megaspore, this megaspore produces the gametophyte. Seeds of *O. muricata* occasionally have two embryos, and the complexes in the twin seedlings indicated that both terminal megaspores in one quartet had developed gametophytes, one with the *curvans* complex from the basal megaspore and the other with the *rigens* complex from the opposite terminal megaspore.

The concept of megaspore competition explained the unusual breeding data for species with one complex in com-

TABLE 5-3

The Effect of Pollen (PL), Egg (EL), and Nonallelic Zygotic (ZL) Lethals on the Types of Progeny from Crosses between Different "Species" of *Oenothera*

Cross

I P_1 *O. suaveolens* (*albicans-flavens*) ♀ × *O. suaveolens* ♂
F_1

	albicans (PL)	*flavens* (ZL-1)
albicans	X	*albicans-flavens*
flavens (ZL-1)	X	X

II P_1 *O. lamarckiana* (*velans-gaudens*) ♀ × *O. lamarckiana* ♂
F_1

	velans (ZL-2)	*gaudens* (ZL-3)
velans (ZL-2)	X	*gaudens-velans*
gaudens (ZL-3)	*gaudens-velans*	X

III P_1 *O. lamarckiana* ♀ × *O. suaveolens* ♂
F_1

	albicans (PL)	*flavens* (ZL-1)
velans (ZL-2)	X	*velans-flavens*
gaudens (ZL-3)	X	*gaudens-flavens*

IV P_1 *O. suaveolens* ♀ × *O. lamarckiana* ♂
F_1

	velans (ZL-2)	*gaudens* (ZL-3)
albicans	*albicans-velans*	*albicans-gaudens*
flavens (ZL-1)	*flavens-velans*	*flavens-gaudens*

V P_1 *O. muricata* (*rigens-curvans*) ♀ × *O. lamarckiana* ♂
F_1

	velans (ZL-2)	*gaudens* (ZL-3)
rigens	*rigens-velans*	*rigens-gaudens*
curvans (EL)	X	X

mon. The *rubens* basal megaspore does not successfully compete with the *velans* terminal megaspore; the *rubens* basal megaspore occasionally competes with the *albicans* terminal megaspore; and, finally, the *rubens* basal megaspore successfully competes with the *curvans* terminal megaspore. In each case, egg lethality could be explained by the developmental competition between megaspores with different complexes at opposite ends of the linear tetrad to produce the gametophyte and not by the inviability of a megaspore or gametophyte. In other words, egg "lethality" is not an intrinsic character but depends on the other complex.

A fortunate observation was responsible for identifying male gametophytes with the *curvans* or *gaudens* complex in the same style. The starch grains in the pollen tubes with the *curvans* complex are subspherical, and those in the pollen tubes with the *gaudens* complex are spindle-shaped. When the pollen from a *curvans-gaudens* hybrid are added to the stigma on the long style of *O. lamarckiana*, the *curvans* pollen tubes elongate more slowly than the *gaudens* pollen tubes. Consequently, all the available eggs are fertilized by the sperm nuclei with the *gaudens* complex. When relatively fewer pollen grains from the *gaudens-curvans* hybrid are added to the stigma on the short style of *O. muricata*, occasional *curvans* sperm nuclei are able to fertilize the still available eggs. This competition between male gametophytes, known as *certation*, has been reported in maize for certain mutant genes (*Ga*) which enhance the growth rate of pollen tubes with the mutant allele.

Cytology

The diploid chromosome number of *O. lamarckiana* and related species is 14. The *semigigas* and *gigas* mutants discovered by de Vries were found to be triploid and tetraploid, respectively, in contrast with most of the other mutants with a somatic chromosome number of 15 ($2n + 1$). While these cytological observations provided significant clues concerning the nature of many mutants, other cytological observations were a mystery which was not readily penetrated. Except for *O. hookeri* with seven bivalents, the related species of *Oenothera* have one or two rings or chains of chromosomes with the same or different numbers of chromosomes in the rings or chains. Furthermore, the rings or chains exhibit a directed orientation at metaphase I. The significance of these observations had to await an understanding of reciprocal translocations and the behavior of interchange complexes at metaphase I (Blakeslee and Cleland, 1930).

Spontaneous breaks in the intercalary segment between the terminal, chiasma-forming segments of the metacentric chromosomes in the *Oenothera* species gave reciprocal translocations. Translocation chromosomes were involved in repeated reciprocal translocations so that complex combinations of translocation chromosomes were produced. While advantageous combinations of genes in different chromosomes were created by pseudolinkage, the sterility usually associated with reciprocal translocations had a negative selective effect. The directed orientation of the interchange complexes, however, ensured that most of the gametophytes would not be deficient. The alternate chromosomes of the interchange complex proceeding to the same pole included the genes responsible for each Renner complex, that is, the two pseudolinkage groups. The exploitation of pollen, egg, and zygotic lethals prevented homozygous complexes, thereby ensuring heterozygosity for the alleles in each complex. Furthermore, the transfer of a monogenically determined phenotype from one Renner complex to another was explained by a crossing-over in the terminal segments of the multiarmed pachytene configuration for the interchange complex. Finally, the relative frequencies of trisomic progeny from the different species were related to the number of chromosomes in the interchange complex; that is, the larger the complex, the greater the probability for an error in the directed orientation of the chromosomes in the complex.

The direct identification of the chromosomes in the different interchange complexes of the *Oenothera* species was not possible. Consequently, a cytological method was devised which assumed that the seven bivalents in *O. hookeri* constituted the arbitrary standards in which the 14 arms could be arbitrarily assigned different numbers: 1·1′ 2·2′ 3·3′ 4·4′ 5·5′ 6·6′ 7·7′. Hybrids between this species and other *Oenothera* species were examined at metaphase I to determine the number of interchange complexes, the number of chromosomes in each complex, and the number of bivalents. For example, the *hookeri-flavens* hybrid has a complex of four chromosomes and five bivalents. The two translocation chromosomes in the *flavens* complex are arbitrarily identified as 1·2′ and 1′·2. The *flavens-velans* combination has two complexes of four chromosomes and three bivalents. Therefore, the *velans* complex has two translocation chromosomes different from those in the *flavens* complex, and so the *velans* translocation chromosomes are arbitrarily assigned 3·4′ and 3′·4. After the analysis of species hybrids, a collection of testers can be assembled to identify the translocation chromosomes in all the species. For example, the *hookeri-franciscana* combination gave one interchange complex of four chromosomes, *velans-franciscana* one complex of six chromosomes, *flavens-franciscana* two complexes of four chromosomes. These observations indicate that the interchange complex in *franciscana* does not include chromosomes 1 or 2 and involves chromosomes 3 or 4 and chromosomes 5, 6, or 7. Eventually, each Renner complex was given a chromosome formula for the organization of the seven chromosomes in terms of chromosome arms. The validity of these formulas was established by successfully predicting chromosome associations in new combinations of complexes which had not been used in earlier crosses.

The *Oenothera* problem confronted cytogeneticists for almost three decades before it was solved. The anomalous breeding behavior of the species and hybrids and the presence of different rings or chains of chromosomes in these species and hybrids finally conformed to the models which had been developed for other animal and plant species. Once the problem was satisfactorily solved, *Oenothera* ceased to be the challenge that had been a major concern of the earlier cytogeneticists.

REFERENCES

ANDERSON, E. G., AND L. F. RANDOLPH. 1945. Location of the centromere on the linkage maps of maize. Genetics 30:518–526.

BLAKESLEE, A. F., AND R. E. CLELAND. 1930. Circle formation in *Datura* and *Oenothera*. Proc. Nat. Acad. Sci. 16:177–183.

BURNHAM, C. R. 1949. Chromosome segregation in maize translocations in relation to crossing over in interstitial segments. Proc. Nat. Acad. Sci. 35:349–356.

———, F. H. WHITE, AND R. LIVERS. 1954. Chromosomal interchanges in barley. Cytologia 19:191–202.

CREIGHTON, H. S., AND B. MCCLINTOCK. 1931. A correlation of cytological and genetical crossing over in *Zea mays*. Proc. Nat. Acad. Sci. 17:492–497.

GARBER, E. D. 1948. A reciprocal translocation in *Sorghum versicolor* Anderss. Amer. J. Bot. 35:295–297.

——— AND T. S. DHILLON. 1961. The genus *Collinsia*. XVIII. A cytogenetic study of radiation-induced reciprocal translocations in *C. heterophylla*. Genetics 47:461–467.

KHUSH, G. S., AND C. M. RICK. 1967. Tomato tertiary trisomics: origin, identification, morphology, and use in determining position of centromeres and arm location of markers. Can. J. Genet. Cytol. 9:610–631.

KIHARA, H., AND M. SHIMOTSUMA. 1967. The use of chromosomal interchanges to test for crossing over and chromosome segregation. Seiken Zoho 19:1–8.

LEWIS, K. R., AND B. JOHN. 1963. Chromosome marker, J. & A. Churchill, London.

MULLER, H. J. 1918. Genetic variability, twin hybrids, and constant hybrids in a case of balanced lethal factors. Genetics 3:422–499.

RENNER, O. 1921. Über Sichtbarwerden des Mendelschen Spaltung im Pollen von *Oenothera*-Bastarden. Ber. Deut. Bot. Ges. 37:129–135.

RHOADES, M. M. 1933. A cytological study of a reciprocal translocation in *Zea*. Proc. Nat. Acad. Sci. 19:1022–1031.

———. 1950. Meiosis in maize. J. Hered. 41:58–67.

———. 1955. The cytogenetics of maize, pp. 123–219. *In*

G. F. Sprague (ed.), Corn and corn improvement. Academic, New York.

Roman, H. 1947. Mitotic nondisjunction in the case of interchanges involving the B-type chromosome in maize. Genetics 32:391–409.

———. 1948. Directed fertilization in maize. Proc. Nat. Acad. Sci. 34:36–42.

——— and A. J. Ullstrop. 1951. The use of A-B translocations to locate genes in maize. Agron. J. 43:450–454.

Soriano, J. D. 1957. The genus *Collinsia*. IV. The cytogenetics of colchicine-induced reciprocal translocations in *C. heterophylla*. Bot. Gaz. 118:139–145.

Stern, C. 1931. Zytologisch-genetische Untersuchungen als Beweise für die Morgansche Theorie des Faktoraustauchs. Biol. Zentralbl. 51:547–587.

Zimmering, S. 1955. A genetic study of segregation in a translocation heterozygote in *Drosophila*. Genetics 40:809–825.

SUPPLEMENTARY REFERENCES

Belling, J. 1914. The mode of inheritance of semi-sterility in the offspring of certain hybrid plants. Z. Induktive Abstammungs- Vererbungslehre 12:303–342.

Burnham, C. R. 1950. Chromosome segregation in translocations involving chromosome 6 in maize. Genetics 35:446–481.

———. 1956. Chromosomal interchanges in plants. Bot. Rev. 22:419–552.

Cleland, R. E. 1962. The cytogenetics of *Oenothera*. Advances Genet. 11:147–237.

Dobzhansky, T., and A. H. Sturtevant. 1931. Translocations between the second and third chromosomes of *Drosophila* and their bearing on *Oenothera*. Carnegie Inst. Washington Pub. 421, pp. 29–59.

McClintock, B. 1930. A cytological demonstration of the location of an interchange between non-homologous chromosomes of *Zea mays*. Proc. Nat. Acad. Sci. 16:791–796.

———. 1931. The order of the genes *C*, *Sh*, and *Wx* in *Zea mays* with reference to a cytologically known point in the chromosome. Proc. Nat. Acad. Sci. 17:485–491.

———. 1945. *Neurospora*. I. Preliminary observations of the chromosomes of *Neurospora crassa*. Amer. J. Bot. 32:671–678.

6 ANEUPLOIDY

Species are characterized by a number of criteria, including their chromosome number and, whenever possible, a karyotype to indicate the morphology of the chromosomes. In plants, the sporophyte has twice as many chromosomes as the gametophytes. In animal species, the male and female can have different chromosome numbers, but the differences are usually related to the number of sex chromosomes. Aneuploidy is the term applied to the loss or gain of one or more chromosomes in sporophyte or gametophyte or in the male or female of an animal species.

The four common types of aneuploidy in diploid species with only bivalents have been rather rigidly defined (Table 6-1). In trisomy ($2n + 1$), an extra (but not supernumerary) chromosome is present, and in monosomy ($2n - 1$), one chromosome is missing. In tetrasomy ($2n + 1^{II}$), the two extra chromosomes are homologous standard chromosomes (a bivalent), and in nullisomy ($2n - 1^{II}$), the missing chromosomes represent a bivalent. The terminology of aneuploidy is also somewhat rigid. For example, the extra chromosome in a

TABLE 6-1

Cytogenetic Characteristics of the Common Types of Aneuploids in Species with Bivalents

Aneuploid Types	Chromosome Number *	Idealized Chromosome Associations †	Genetic Characteristics
Monosomy	$2n - 1$	IIs + 1^{I}	Hemizygous for one linkage group
Nullisomy	$2n - 1^{II}$	IIs	Homozygous deficiency for one linkage group
Trisomy	$2n + 1^{I}$	Primary: IIs + 1^{III}	Three alleles for loci in one linkage group
		Secondary: IIs + 1^{III}	Four alleles for loci in one arm of one linkage group
		Tertiary IIs + 1^{V}	Three alleles for loci in segments of two different linkage groups
Tetrasomy	$2n + 1^{II}$	IIs + 1^{IV}	Four alleles for loci in one linkage group

* The roman numeral refers to a pair of homologous chromosomes.

† The roman numeral refers to the number of chromosomes in an association at diakinesis or metaphase I.

trisomic is a *trisome,* and the one chromosome in a monosomic is a *monosome.*

Trisomy and monosomy are more common than tetrasomy or nullisomy. During the numerous mitotic divisions from zygote to mature individual, the occasional nondisjunction of sister chromosomes at anaphase yields one trisomic and one monosomic cell. The aneuploid cells initiate a clone of trisomic or monosomic cells which are included in different organs or tissues, depending on their ability to compete with the standard (euploid) cells. In basic diploid species, the trisomic rather than the monosomic cells are likely to produce an aneuploid clone in the germinal tissue, and the spores or gametes will have either the haploid (n) or disomic ($n + 1$) chromosome number. The frequencies of these two types of spores or gametes are determined by the extent of the chromosomal mosaicism in the germinal tissue. Chromosome nondisjunction can also occur at either meiotic division to give haploid or aneuploid spores or gametes. Mitotic or meiotic nondisjunction in polyploid species with bivalents can produce functional monosomic or disomic gametophytes.

MONOSOMY

The loss of a standard chromosome is a major deficiency which usually leads to inviable or malfunctioning gametophytes or to lethality for the zygote or embryo. The loss of a sex chromosome in the gametes of animal species does not necessarily produce an inviable zygote or embryo as long as one X chromosome is present. One interesting exception to this generality has been found in *Drosophila melanogaster,* where the loss of the very small chromosome IV can be tolerated. Although monosomics have been found in a few diploid species, monosomy is generally restricted to tetraploid or hexaploid species.

In tetraploid plant species, the four sets of chromosomes represent two sets from one parental species and two sets from another parental species. The spores contain two sets, one from each of the parental species which were closely related and presumably shared numerous homologous genes. Consequently, tetraploid species have numerous duplications. The loss of a chromosome in spores from a polyploid species is not necessarily lethal because the duplication presumably compensates for the deficiency. As a rule, the female gametophyte tolerates a deficiency more frequently than the male gametophyte in that the deficiency is less likely to cause inviability or malfunction. Consequently, the cytogenetics of monosomy has been almost completely restricted to tetraploid or hexaploid species with bivalents.

Monosomics are occasionally detected in the progeny of a polyploid species by an altered phenotype which must be confirmed by a chromosome count. The frequency of monosomic progeny can be increased by subjecting young embryos to temperature shocks or nonlethal dosages of ionizing radiation, which cause mitotic nondisjunction and eventually some germinal cells with a missing chromosome. A more efficient method utilizes an asynaptic or desynaptic mutant or a variety with numerous univalents at metaphase I; progeny from crosses between standard and mutant parents include numerous monosomics detectable by their altered phenotype or chromosome count.

During meiosis in a monosomic, the homologous chromosomes form bivalents while the monosome appears as a univalent. If the monosome were to proceed through meiosis in a normal manner, two of the four spores would be haploid (n) and the other two spores aneuploid ($n - 1$). If the monosome lags during the first meiotic division and is lost in the cytoplasm, all four spores will be aneuploid. The two chromatids of the monosome may or may not separate at anaphase I, and the sister chromosomes may or may not be included in a spore nucleus at telophase II. For these reasons the frequency of $n - 1$ spores usually exceeds 50 percent (Table 6-2).

The impact of a missing chromosome on gametophyte viability, the proper functioning of the male gametophyte to effect fertilization, zygote viability and the phenotype of the sporophyte depends on the number of chromosome sets in the species and, for a tetraploid species, on the particular missing chromosome. A comparison of the reproductive features of the different monosomes in the tetraploid species *Nicotiana tabacum* (Table 6-3) and in the hexaploid species *Triticum aestivum* (Table 6-4) illustrates these effects in species with four or six sets of chromosomes, respectively. The tetraploid species does not yield nullisomic progeny.

Monogenically determined phenotypes are less common in *N. tabacum* than in diploid species, and duplicate genes (15:1 ratio) are readily obtained in this species. The large number of chromosomes and the relative paucity of monogenic characters were responsible for the slow progress in construct-

TABLE 6-2

The Number of Micronuclei in Quartets of Microspores from Four Monosomic Plants of *Nicotiana tabacum* *

Monosomic	Number of Micronuclei: 0	1	2	3+	$n - 1$ Spores, %
A	322	279	394	11	76.8
B	226	126	240	16	75.6
N	125	88	150	21	76.7
R	180	121	181	8	75.0
Chromosome constitution	$n \mid n$ / $n-1 \mid n-1$	$n \mid n-1$ / $n-1 \mid n-1$	$n-1 \mid n-1$ / $n-1 \mid n-1$		

* Quartets with three or more micronuclei presumably represent more than one univalent in the monosomics.

Source: Olmo, 1935.

TABLE 6-3

Ovular Transmission of Monosomes in *Nicotiana tabacum*

Monosome	Total	Monosomics	
		No.	%
A	1,291	1,016	78.7
B	1,303	421	32.3
C	1,331	610	45.8
D	835	344	41.2
E	819	671	81.9
F	879	526	59.8
G	972	62	6.4
H	888	625	70.4
I	520	40	7.7
J	432	26	6.0
K	381	184	48.3
L	183	34	18.6
M	803	480	59.8
N	784	173	22.1
O	1,085	838	77.2
P	1,049	68	6.5
Q	845	93	11.0
R	1,074	584	54.4
S	1,093	339	31.0
T	437	48	11.0
U	231	84	36.4
V	242	19	7.9
W	234	12	5.1
Z	223	18	8.1

Source: Clausen and Cameron, 1944.

ing linkage groups and for assigning the groups to the corresponding chromosomes. Clausen and Cameron (1944) successfully exploited monosomy in this species to assign genes to specific chromosomes by pseudodominance (hemizygosity) for mutant alleles on the monosome. Hemizygosity normally occurs in animal species where genes on the X chromosome do not

TABLE 6-4

Frequencies of Nullisomics in Progenies from Self-pollinated Monosomics of *Triticum aestivum*

Monosome	Number of Plants	Nullisomic No.	Nullisomic %	Numbers Cytologically Confirmed
I	671	16	2.4	10
II	962	48	5.0	26
III	2,682	205	7.6	84
IV	2,159	138	6.4	52
V	1,431	14	1.0	5
VI	598	15	2.5	5
VII	885	12	1.4	10
VIII	192	7	3.6	3
IX	832	28	3.4	1
X	1,457	13	0.9	6
XI	331	11	3.3	8
XII	125	3	2.4	3
XIII	255	6	2.4	0
XIV	381	8	2.1	6
XV	1,924	113	5.9	38
XVI	712	35	5.8	17
XVII	1,109	27	2.4	17
XVIII	575	5	0.9	2
XIX	1,084	30	2.8	11
XX	572	25	4.4	9
XXI	177	2	1.1	2

Source: Sears, 1954.

have alleles on the Y chromosome. In *N. tabacum*, the recessive allele (*wh*) for white flower was assigned to chromosome C by crossing a monosomic seed parent with the dominant allele (*Wh*) for carmine flower on chromosome C and a standard recessive (*wh wh*) plant and detecting recessive progeny. The frequency of the recessive progeny is not important in assigning the gene to chromosome C.

The number of different monosomics for a species is equal to the gametophyte chromosome number. Thus, *N. tabacum* ($n = 24$) will have 24 different monosomics. The few spontaneous monosomics found in this species provided the initial experimental material for demonstrating the value of monosomy in locating genes in particular chromosomes. The discovery of an asynaptic race was responsible for the eventual isolation of all 24 monosomics. By crossing asynaptic plants with dominant alleles and standard plants homozygous for a number of recessive alleles, the recessive progeny not only indicated monosomy for these phenotypes but also related a particular gene with a specific monosome. While the 24 different monosomics might have characteristically altered phenotypes, the available pool of recessive mutations is the limiting factor in isolating and detecting each monosomic. Genes assigned to the same monosome obviously constitute a linkage group, and appropriate crosses for these genes provide the necessary data for constructing a linkage map. Since the progeny have bivalents, the sterility of the monosomics is not caused by asynapsis.

The parental diploid species of the tetraploid species *N. tabacum* are assumed to be *N. tomentosiformis* and N. *sylvestris* or closely related species which have become extinct. The yellow-burley phenotype in *N. tabacum* is determined by duplicate genes (*yb-1*, *yb-2*), and each gene has been placed in different chromosomes, using monosomic analysis (Table 6-5). Different monosomics with green leaves were crossed with standard recessive plants and the standard or monosomic green progeny were identified by morphology, sterility, and chromosome number. These progeny were self-pollinated and crossed with standard recessive pollen parents. The breeding data indicated that the *yb-1* locus was on chromosome B and the *yb-2* locus on chromosome O. One chromosome is presumably contributed by *N. sylvestris* and the other by *N. tomentosiformis*.

TABLE 6-5

The Inheritance of the Duplicate Genes (*yb-1 yb-2*) for the Yellow-burley Phenotype in a Monosomic of *Nicotiana tabacum*

P_1	Monosomic green ×	Diploid yellow burley
	B^{+} O^{+} O^{+}	$B^{yb\text{-}1}$ $B^{yb\text{-}1}$ $O^{yb\text{-}2}$ $O^{yb\text{-}2}$
F_1	Diploid green	Monosomic green
	B^{+} $B^{yb\text{-}1}$ O^{+} $O^{yb\text{-}2}$	$B^{yb\text{-}1}$ O^{+} $O^{yb\text{-}2}$

Progenies

Phenotype	No.	Phenotype	No. $2n$	No. $2n - 1$
Self-pollination:				
Green	79	Green	25	41
Yellow burley	6	Yellow burley	11	12
7% mutants		35.9% mutants		
Testcross:				
Green	64	Green	33	10
Yellow burley	20	Yellow burley	19	19
23.8% mutants		46.9% mutants		

Source: After Clausen and Cameron, 1944.

NULLISOMY

In *N. tabacum* the $n - 1$ pollen grains are normal, abnormal, or aborted in appearance. The apparently normal pollen grains, however, rarely function. Consequently, self-pollinated monosomics in this species yield standard or monosomic but not nullisomic progeny. In the hexaploid species *Triticum aestivum*, self-pollinated monosomics give standard, monosomic, or nullisomic progeny (Table 6-4). Although the $n - 1$ pollen grains can function to effect fertilization, the frequencies of functional $n - 1$ pollen are determined by several factors, particularly the impact of the specific missing chromosome on the normal development and growth of the pollen tube, and range from 1 to 19 percent. The frequencies of functional $n - 1$ eggs with different missing chromosomes range from 61 to 86 percent. The mean frequency for the different types of functional $n - 1$ pollen grains is 4 percent and for different types of functional $n - 1$ eggs, 75 percent. The frequencies of standard monosomic and nullisomic progeny from a self-pollinated monosomic in *T. aestivum* are readily calculated (Fig. 6-1). While approximately half of the 21 nullisomics are either male-sterile or female-sterile, the progenies for the corresponding fertile monosomics include 0.9 to 7.6 percent nullisomics which can be used in crosses as the seed parent.

Although the monosomics of *T. aestivum* do not have markedly altered phenotypes, each of the 21 nullisomics is readily distinguished. A comparison of the 21 nullisomic phenotypes indicated that three nullisomics were similar and seven groups were present. By appropriate crosses, plants were obtained which were nullisomic for one chromosome and tetrasomic for another chromosome. Plants with certain nullisomic-tetrasomic combinations closely resembled the standard diploids, while other combinations were obviously different from the standard diploids. The effective combinations indicated that certain tetrasomes could compensate for the absence of

Eggs	Pollen	
	21 chromosomes (n) 96% (81–99)	20 chromosomes (n−1) 4% (1–19)
21 chromosomes (n) 25% (14–29)	Disomic (2n) 24% (11–29)	Monosomic ($2n-1^{II}$) 1% (0.1–5)
20 chromosomes (n−1) 75% (61–86)	Monosomic (2n−1) 72% (49–85)	Nullisomic ($2n-1^{II}$) 3% (0.6–16)

Fig. 6–1. Types and frequencies of progeny from self-pollinating a monosomic ($2n - 1$) wheat plant. The average frequencies and ranges (*parentheses*) are calculations based on experimental observations. [*R. Morris and E. R. Sears. 1967. The cytogenetics of wheat and its relatives, pp. 19–87. In K. S. Quisenberry and L. P. Reitz (eds.), Wheat and wheat improvement. American Society of Agronomy Monograph 13.*]

particular disomes. These observations were interpreted by assuming that the three chromosomes in each of the seven groups have considerable homologous genetic information even though these chromosomes are not identical. Such chromosomes are termed *homoeologous* rather than homologous. Therefore, *T. aestivum* has seven groups, each containing three homoeologous chromosomes. Each of three basic diploid species with seven pairs of chromosomes involved in the origin of the hexaploid species with 21 pairs of chromosomes furnished one of the three homoeologous chromosomes in each group.

Attempts to assign mutant genes to specific chromosomes in *T. aestivum* were frustrated by the relatively small stockpile of monogenically determined phenotypes and the relatively high chromosome number in this important crop species. An extensive genetic analysis was not feasible until nullisomic analysis was successfully exploited by E. R. Sears, who first isolated and identified the 21 different nullisomics in the Chinese Spring variety. This work has served as the model for similar studies in other varieties of *T. aestivum* and in other polyploid species.

Sears (1953) developed different methods for assigning mutant genes to specific chromosomes by nullisomic analysis. The complex genetic constitution of this species resulting from the duplications among homoeologous chromosomes precluded the application of a simple generalized method. All the methods, however, require a complete set of different monosomics or nullisomics.

Nullisomics from a standard plant with a dominant phenotype can exhibit a recessive phenotype. For example, seeds of the Chinese Spring variety are red, but the seeds from only nullisomic XVI (lacking the bivalent for chromosome XVI) of the 21 different nullisomics are white. In this case, the homozygous deficiency is equivalent to the homozygous recessive mutation for white seeds. A similar example occurs in maize, where a homozygous deficiency at the brown-midrib (*bm-1*) locus yields the recessive phenotype. Not all recessive phenotypes, however, are expressed by each nullisomic. Al-

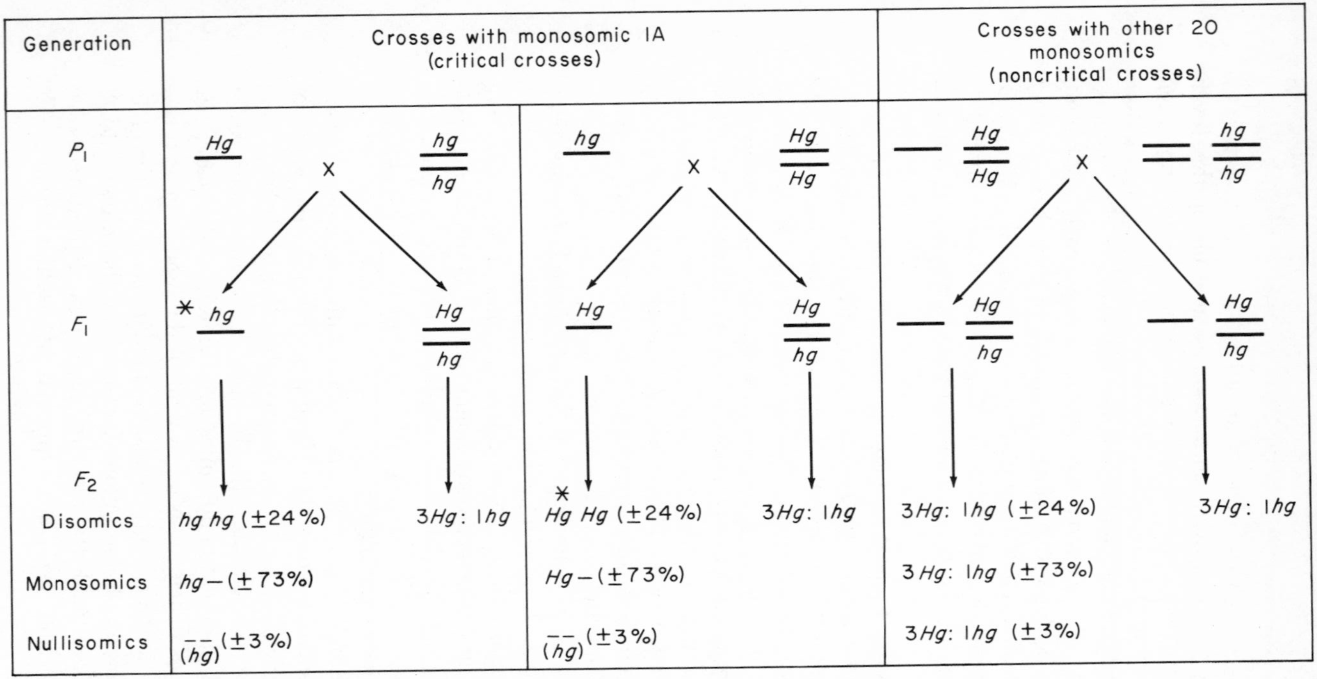

Generation	Crosses with monosomic IA (critical crosses)				Crosses with other 20 monosomics (noncritical crosses)		
P_1	*Hg* —	×	*hg*/*hg*	*hg* — × *Hg*/*Hg*	— *Hg*/*Hg*	×	= *hg*/*hg*
F_1	* *hg* —		*Hg*/*hg*	*Hg* — ; *Hg*/*hg*	— *Hg*/*hg*		— *Hg*/*hg*
F_2							
Disomics	*hg hg* (±24%)		3*Hg*: 1*hg*	* *Hg Hg* (±24%) ; 3*Hg*: 1*hg*	3*Hg*: 1*hg* (±24%)		3*Hg*: 1*hg*
Monosomics	*hg*—(±73%)			*Hg*—(±73%)	3*Hg*: 1*hg* (±73%)		
Nullisomics	——/(*hg*) (±3%)			——/(*hg*) (±3%)	3*Hg*: 1*hg* (±3%)		

Fig. 6.2. The assignment of genes to specific chromosomes in wheat by the phenotypes and their frequencies in the F_1 and F_2 progeny from crosses involving the 21 monosomics as the female; * denotes progeny which determines chromosomal location of gene. (*R. Morris and E. R. Sears. 1967. The cytogenetics of wheat and its relatives, pp. 19–87, fig. 10, p. 55. In K. S. Quisenberry and L. P. Reitz (eds.), Wheat and wheat improvement. American Society of Agronomy Monograph 13.*)

though the recessive mutation for virescent seedlings in this variety is in chromosome III, nullisomic III seedlings are not virescent.

The second method requires a cross between each of the 21 nullisomics with the dominant phenotype and a standard recessive plant, followed by self-pollination of the monosomic or standard hybrids to obtain segregating F_2 populations (Fig. 6-2). The absence of recessive hybrid monosomics is explained by stipulating that the recessive allele is hemizygous-ineffective but homozygous-effective. In the critical hybrid monosomic, all the standard progeny are recessive, in contrast with dominant and recessive progeny from the other 20 hybrid monosomics. For some phenotypes, such as disease resistance, a relatively large population is needed for conducting extensive tests. In these cases, the critical hybrid monosomic is self-pollinated to produce recessive standard plants with high fertility. These plants are self-pollinated to produce numerous F_3 progeny for the tests to detect the phenotype in question.

The third, or chromosome-substitution, method has interesting implications other than assigning mutant genes to particular chromosomes. One variety of related species may be resistant to a disease but lack other desirable characteristics present in a second variety. The development of a new variety with disease resistance and other desirable characteristics by intervarietal or interspecific hybridization can be a lengthy and tedious process. In the chromosome-substitution method, the standard donor variety is crossed with each of the 21 monosomics in the recipient variety. The monosomic hybrids are then crossed with the standard recipient, and the monosomic progeny in each generation are tested and selected for crossing with the standard recipient. By the fifth generation of crossing selected monosomic hybrids with the recurrent recipient, approximately 97 percent of the genotype contributed by the noncurrent donor has been eliminated, except for the monosome which was protected from crossing-over. The monosomic hybrids at the fifth or later generation are self-pollinated, and the fertile standard progeny with bivalents have one homozygous bivalent from the donor (Fig. 6-3). Chromosome substitution also has been used to replace one bivalent in *T. aestivum* with one bivalent from another species.

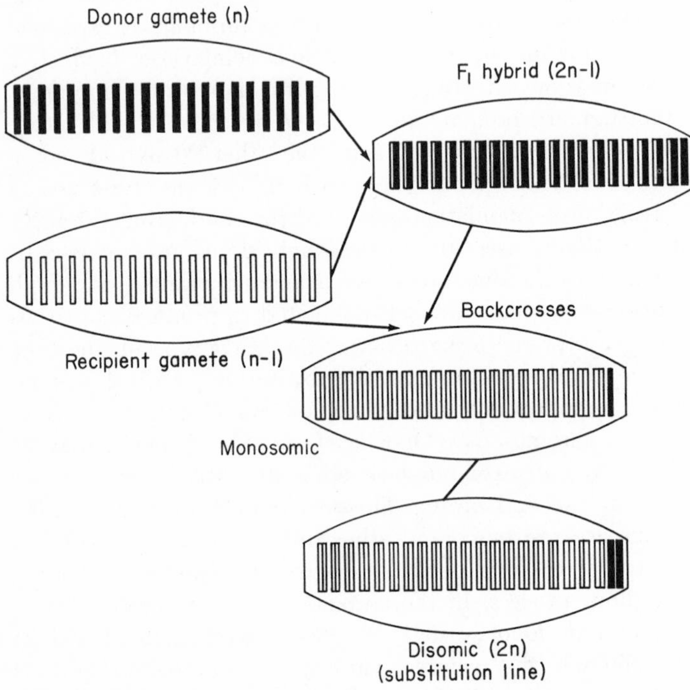

Fig. 6–3. Development of a substitution line in which one pair of chromosomes from a donor variety replaces the identical pair in the recipient variety. The repeated backcrosses using the monosomic in each generation are responsible for the recovery of the genotype of the recipient variety for the remaining chromosomes. In wheat, this method can be used to obtain the 21 substitution lines. [*R. Morris and E. R. Sears. 1967. The cytogenetics of wheat and its relatives, pp. 19–87, fig. 4, p. 32. In K. S. Quisenberry and L. P. Reitz (eds.), Wheat and wheat improvement. American Society of Agronomy Monograph 13.*]

TRISOMY

Trisomic individuals are detected by their altered phenotype, particularly in diploid species, or by their chromosome number ($2n + 1$). The extra chromosome is either standard or derived from a standard chromosome. In animal species, a distinction is made between the sex chromosomes and the autosomes in terms of trisomy. The constitution of the extra chromosome is responsible for primary, secondary, or tertiary trisomy (Table 6-1). Two additional types will be discussed later as telocentric and compensating trisomy.

The extra chromosome in a primary trisomic is a standard chromosome, in a secondary trisomic an isochromosome, and in a tertiary trisomic a translocation chromosome (Fig. 6-4). These categories are generally distinguished by their characteristic phenotypes, breeding data for gene markers, and certain diagnostic chromosome associations at diakinesis or metaphase I. In the primary trisomic, the three homologous chromosomes usually form a chain, but other configurations can occur, depending on the number of chiasmata in each arm. A ring of three chromosomes or a chiasma in the homologous arms of the univalent characterizes the secondary trisomic. Finally, a chain of five chromosomes indicates a tertiary trisomy. The translocation chromosome serves as a bridging chromosome with homologous segments for different chromosomes.

The additional segments in the trisome are responsible for duplicated chromosomal (and presumably genetic) material in a diploid species. A primary trisomic has three alleles for loci in the trisome; a secondary trisomic has four alleles for loci in the arms of the isochromosome; and a tertiary trisomic has three alleles for loci in each segment of the translocation chromosome. The maximum number of different primary trisomics in a diploid species is equal to the gametophyte or gametic chromosome number and the maximum number of different secondary trisomics, twice the gametophyte or gametic chromosome number. The number of different tertiary

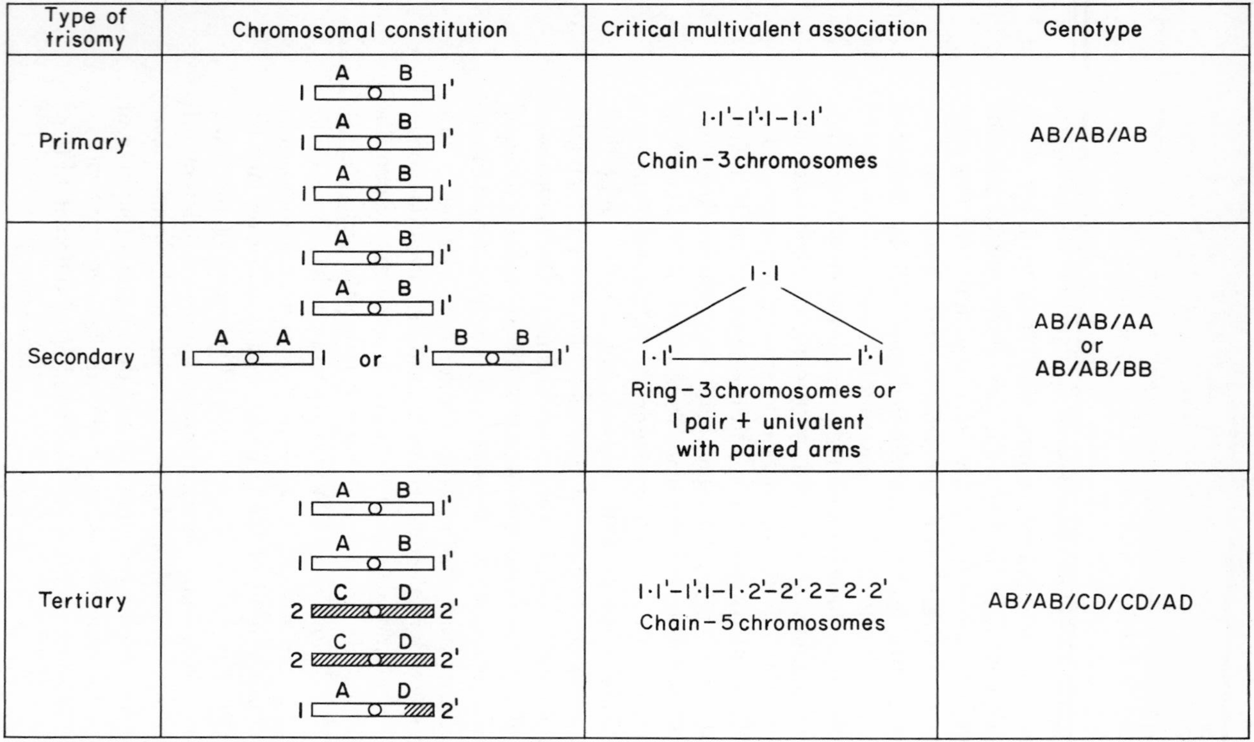

Type of trisomy	Chromosomal constitution	Critical multivalent association	Genotype
Primary	1 A ○ B 1' 1 A ○ B 1' 1 A ○ B 1'	1·1'–1'·1–1·1' Chain–3 chromosomes	AB/AB/AB
Secondary	1 A ○ B 1' 1 A ○ B 1' 1 A ○ A 1 or 1' B ○ B 1'	1·1 1·1'——1'·1 Ring–3 chromosomes or 1 pair + univalent with paired arms	AB/AB/AA or AB/AB/BB
Tertiary	1 A ○ B 1' 1 A ○ B 1' 2 C ○ D 2' 2 C ○ D 2' 1 A ○ D 2'	1·1'–1'·1–1·2'–2'·2–2·2' Chain–5 chromosomes	AB/AB/CD/CD/AD

Fig. 6–4. The chromosomal constitution, critical multivalent associations at diakinesis or metaphase I, and genotype in the primary, secondary, and tertiary trisomics of a diploid species.

trisomics is limited only by the number of reciprocal translocations in the species.

The phenotypic impact of duplications for the different types of trisomics cannot be predicted for any diploid species and must be established empirically. Different primary trisomes can be expected to produce characteristic phenotypic alterations because each chromosome has a specific aggregation of genes (linkage group). In relating a characteristic phenotypic alteration with a specific primary trisome, the standard and primary trisomics are raised under different environmental conditions, such as the field, greenhouse, or controlled-environment facility. In some species, all the primary trisomics can be distinguished from each other and from the diploids, and in other species, only some primary trisomics can be distinguished. The cytogenetics of trisomy was first thoroughly investigated in the plant species *Datura stramonium* by Blakeslee (Avery, Satina, and Rietsema, 1959). Although the chromosomes of this species are not readily distinguished from each other, the plants are extremely sensitive to the duplications and exhibit characteristically altered phenotypes, depending on the type of trisome. Rick and collaborators have obtained different types of trisomics in the tomato (*Lycopersicon esculentum*, $n = 12$), a species with excellent pachytene chromosomes, comprehensive linkage groups, and distinctive phenotypic alterations caused by duplications. The cytogenetic consequences of trisomy in plant and animal species will be illustrated with data from *Datura stramonium, L. esculentum,* and *Drosophila melanogaster.*

Primary Trisomy

Primary trisomics occur spontaneously in relatively low frequencies in the progenies of standard individuals. A reasonably accurate estimate of these frequencies requires a species in which each primary trisome is responsible for a readily detected and characteristic phenotype. A population of 125,027 plants of *Datura stramonium* ($n = 12$) had a total of 775 primary trisomics, and each trisome had a characteristic frequency. The different frequencies are attributable to at least three factors: (1) the frequency of mitotic or meiotic nondisjunction for each chromosome, (2) the impact of each chromosome on the viability or functioning $n + 1$ gametophytes, and (3) its impact on the viability or development of trisomic zygotes or embryos. Relatively high frequencies of primary trisomics are found in progenies from fertile autotriploids ($3n$) or asynaptic or desynaptic mutants.

The first known primary trisomic was found in *Datura stramonium* as a mutant (Globe) with relatively broad leaves and a globose seed capsule. The breeding data indicated that the inheritance of this mutation did not fit an orthodox Mendelian ratio for a monohybrid (Table 6-6). A cytological examination of the mutants revealed the constant presence of an extra chromosome, a primary trisome. The other 11 primary trisomes were eventually found, and each was associated with characteristic alterations in the morphology of the seed capsule, leaves, and other organs. While primary trisomics have been obtained for numerous plant species, relatively few animal species have yielded primary trisomics. It is interesting to note that man has provided more examples of primary trisomy than any other animal species. In some plant species, not all the primary trisomics exhibit an obviously altered phenotype and must be detected by their chromosome number.

TABLE 6-6

Inheritance of the Globe Mutant in *Datura stramonium*

Cross		Seeds	%	Progeny		
					Globe	
♀	♂	Planted	Seedlings	Normal	No.	%
Globe	Globe	3,015	59.1	1,382	400	22.5
Globe	Normal	2,628	73.6	1,435	500	25.8
Normal	Globe	2,771	65.4	1,759	53	2.9

Source: Blakeslee, 1921.

Considerable data have been accumulated on the transmission of the primary trisomes by the male or female gametophyte; relatively little information is available on the transmission of primary trisomes by the gametes of animal species. As a rule, the extra chromosome is more frequently transmitted by the female than the male gametophyte (Table 6-7) and, in some species, almost exclusively by the female gametophyte. Progeny of self-pollinated trisomics of a diploid or tetraploid species generally include diploids and trisomics; of a hexaploid species, generally diploids, trisomics, and tetrasomics. Attempts to formulate other generalizations to account for the transmission of primary trisomes have not been meaningful. For example, transmission frequencies seem to be related to chromosome length in maize but not in *Datura stramonium*. Since each chromosome in a diploid species has a characteristic ensemble of genes and the genic imbalance in

TABLE 6-7

Frequencies of Transmission of the Trisomes of Tomato, $n = 12$

	Frequency in Progenies					
	$2n + 1$ ♀ × $2n$ ♂			$2n + 1$: Self-pollinated		
Chromosome	Total	$2n + 1$	$2n + 1$, %	Total	$2n + 1$	$2n + 1$, %
1	439	20	4.6			
2	409	18	4.4	193	16	8.3
3	661	7	1.1			
4	311	77	24.7	467	146	31.3
5	598	133	22.2	344	88	25.6
6	224	1	0.4			
7	593	88	14.8	294	53	18.0
8	717	156	21.7	772	193	25.0
9 *	517	86	16.6	423	70	16.5
10	666	133	20.0	237	55	23.2
11	415	61	14.6	627	116	18.5
12	476	75	15.7	300	52	17.3

* Numbers of trisomics do not include a large proportion of triplo-9 seedlings that lack growing points.

Source: Rick and Barton, 1954.

the $n + 1$ gametophyte or primary trisomic is associated with this ensemble, transmission frequencies are related to the impact of the imbalance on the viability or functioning of the gametophyte, zygote, or embryo.

The behavior of the primary trisome during the meiotic divisions is similar to that of a univalent. Consequently, primary trisomics are likely to produce more n than $n + 1$ gametophytes or gametes. In plants, $n + 1$ pollen grains abort, fail to germinate, swell or burst on the stigma, or produce slow-growing pollen tubes. For example, the transmission of primary trisome 6 in *Datura stramonium* was 9.2 percent for routinely accomplished crosses. This transmission frequently could be manipulated by using relatively few pollen grains and separating the seeds in the upper or lower half of the seed capsule (Table 6-8). The $n + 1$ pollen grains produce a slow-growing tube which cannot compete with the tubes from haploid pollen grains in fertilizing the available eggs. After the pollen grains germinated and some had fertilized the eggs, the styles were cut near the base. The absence of trisomic seeds indicated

TABLE 6-8

Frequency of Primary Trisomics in Progeny from Crosses Using a Primary Trisomic (Number 6) Pollen Parent and a Diploid Seed Parent in *Datura stramonium*

	Progeny		
		Trisomic	
Treatment	Total	No.	%
Standard pollination:			
Entire seed capsule	551	57	9.2
Sparingly pollinated:			
Upper seed capsule	502	7	1.4
Lower seed capsule	580	299	56.4
Timed excision of style:			
Entire seed capsule	116	0	0.0

Source: Buchholz and Blakeslee, 1930.

that the slow-growing tubes were removed with the excised styles. Viable trisomic embryos are often less vigorous than diploid embryos, so that the routine planting or growing procedures favor the survival of diploid progeny. For example, progeny from one primary trisomic in *Datura stramonium* included 12 percent trisomics when the seeds were routinely planted and gave 69 percent trisomics when the smallest seeds were carefully planted and the seedlings were spaced.

Genetic Studies

The first cytogenetic analysis of a trisomic was accomplished in *Drosophila melanogaster* and provided convincing evidence that the genes are indeed in the chromosomes (Bridges, 1916). Because primary trisomy is restricted to an extra standard chromosome, trisomic flies (XXY) with an extra sex chromosome are not properly classified as a primary trisomic. Genetically, the Y chromosome is treated as though it had only recessive alleles for loci on the X chromosome. The first trisomic was detected in the progeny from a cross involving a recessive sex-linked mutant. The recessive females and dominant males in progeny from a cross between a recessive female and a dominant male were explained by assuming a meiotic nondisjunction of the X chromosomes in the female parent (Fig. 6-5). The dominant males were sterile, and the recessive females were fertile. Breeding data from crosses between these females and standard males indicated that the exceptional females were XXY, which was confirmed cytologically; the exceptional sterile males were XO, and their sterility resulted from the nonmotility of their sperm. Primary trisomic (XXX) females usually died at the pupal stage. The initial event responsible for eggs with no X or XX chromosomes has been termed a *primary nondisjunction,* the distribution of the XXY chromosomes in the exceptional fertile females being called *secondary nondisjunction.* The breeding data from crosses between recessive XXY females and dominant stand-

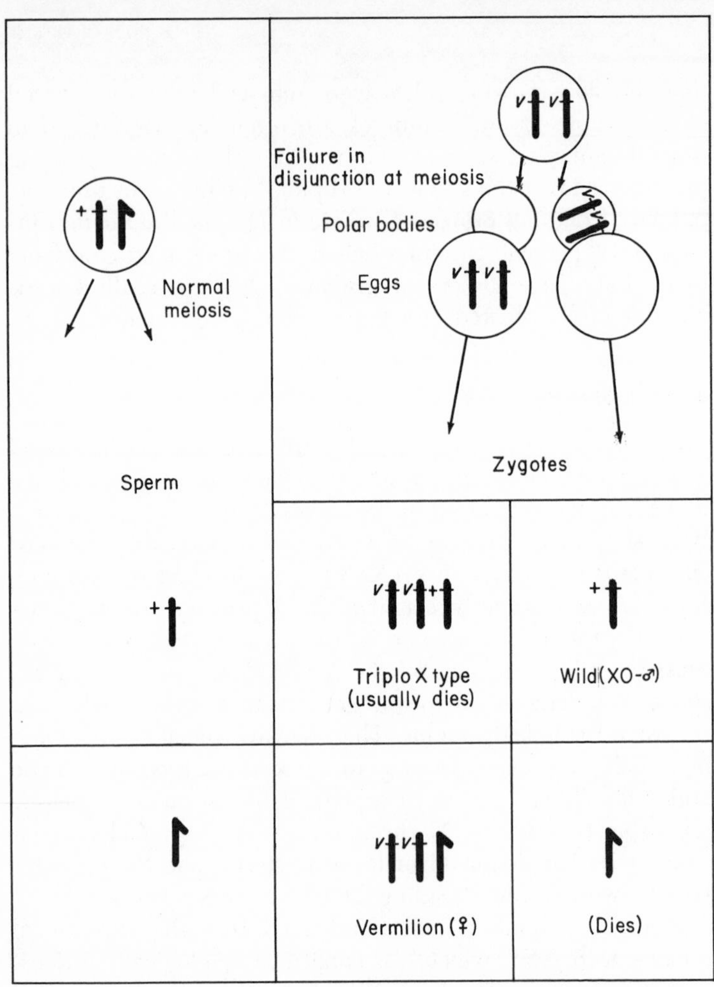

Fig. 6–5. Diagram illustrating the primary nondisjunction of the X chromosome of *Drosophila melanogaster* during oogenesis to yield eggs with both or with no X chromosomes and their detection by the unexpected genotype of certain progeny. The two meiotic divisions are not presented. (*After Bridges, 1916.*)

ard males differ from the results for the usual sex-linked inheritance. By assuming that the recessive allele was indeed in the X chromosome, Bridges predicted not only the chromosome number ($2n$ or $2n + 1$) but also the type of extra chromosome in trisomic recessive or dominant progeny. The evidence that genes might be in the chromosomes had been circumstantial and persuasive but not conclusive until Bridges published the results of his study on the cytogenetics of trisomy in *Drosophila melanogaster*.

The frequencies of the different phenotypes in each set are explained by assuming the preferential pairing of the structurally identical X chromosomes in approximately 92 percent of the oocytes of the trisomic female (Table 6-9). In these oocytes, the X chromosomes proceed to opposite poles at anaphase I and the Y chromosome to either pole, yielding X or XY eggs. In approximately 8 percent of the oocytes, one X and the Y chromosome go to opposite poles and the other X chromosome to either pole, giving eggs with XY or XX. The concept of a preferential pairing of chromosomes was an innovation which had considerable influence on the construction of cytogenetic models to account for unexpected breeding data.

Bridges observed 18 cases of "equational exceptions" (nondisjunction at the second meiotic division) in progenies from crosses between females with two recessive sex-linked genes and males with dominant alleles. The exceptional females had one or the other recessive phenotype. In 15 cases, the females were XXY, and in 3, were XX. Breeding data for 12 equational exceptions indicated that one X chromosome had been involved in a crossing-over and the other X chromosome was a nonrecombinant (Fig. 6-6). According to Bridges,

> XX synapsis took place; each X split so that a four strand stage occurred; crossing over took place between two only of these strands, one from each X; the reduction division separated the paternal X from the maternal X, each cell receiving a non-crossover and at the same time a crossover strand; at the next division, these two strands ordinarily enter different cells, but by an occasional nondisjunction these

TABLE 6-9

Inheritance of the White-eye (*w*) Mutant in XXY Females of *Drosophila melanogaster* in Crosses with the Wild-type Red-eye (+) Males

A. Data

Parents	Red-eye (Normal) Male	
White-eye female: abnormal stock		2.1% of white-eye females, all with some exceptional progeny
	2.2% of red-eye males, all with normal progeny	
	49.0% of red-eye females, half with normal progeny, and half with some exceptional progeny	46.7% of white-eye males, half with normal progeny, and half with some exceptional progeny

B. Interpretation

Parents				XY (Red-eye Male) Sperm: X^+		XY (Red-eye Male) Sperm: Y	
X^wX^wY (white-eye female: abnormal stock)	Eggs	8.2%	X^wX^w	$X^+X^wX^w$	Dies	X^wX^wY	w ♀
			Y	X^+Y	+ ♂	YY	Dies
		91.8%	X^w	X^+X^w	+ ♀	X^wY	w ♂
			X^wY	X^+X^wY	+ ♀	X^wYY	w ♂

Source: After Bridges, 1916.

strands do not separate from each other at the equational division and consequently enter the same nucleus. In the case of an XX female the presence of the Y might favor the process by entering the other cell so that one cell receives two X chromosomes and the other 2 Y's. Equational nondisjunction thus enables us to examine at leisure the products of reduction.

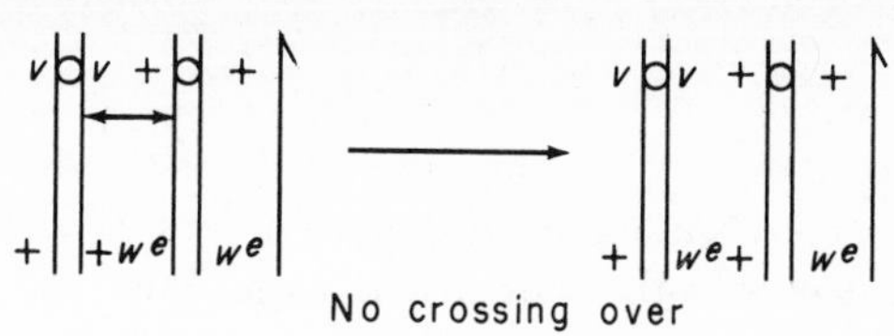

No crossing over

Sperm

Eggs	++		Y	
n+1	Genotype	Phenotype	Genotype	Phenotype
v +/Y	v +/+ +/Y	+♀	v +/Y/Y	v ♂
+ w^e/Y	+ w^e/+ +/Y	+♀	+ w^e/Y/Y	w^e♂
+ w^e/v +	+ w^e/v +/+ +	Usually dies	+ w^e/v +/Y	+ ♀
n				
v +	v +/+ +	+♀	v +/Y	v♂
+ w^e	+ w^e/+ +	+♀	+ w^e/Y	w^e♂
Y	+ +/Y	+♂	YY	Dies

Crossing over-meiotic nondisjunction

Exceptional progeny

n+1				
v +/v w^e	v +/v w^e/+ +	Usually dies	v +/v w^e/Y	v♀
+ w^e/v w^e	+ w^e/v w^e/+ +	Usually dies	+ w^e/v w^e/Y	w^e♀

v +, + w^e = parental chromatids; v w^e = crossover chromatid.

Fig. 6–6. Origin of exceptional progeny from a female (XXY) *Drosophila melanogaster* heterozygous for two sex-linked mutant genes in repulsion. Genotypes of exceptional progeny are determined by breeding the females. (*After Bridges, 1916.*)

The equational exceptions provided the first evidence that crossing-over between homologous chromosomes, whatever the mechanism, occurs at the four-strand stage. Furthermore, meiotic nondisjunction gave two of the four strands from the one meiocyte so that deductions could be drawn concerning events resulting from a crossing-over. One advantage in conducting genetic studies with a number of algal or fungal species has been the recovery of all the products from one meiocyte to detect recombinant or parental chromosomes.

Although the XXY females in *D. melanogaster* provided acceptable evidence that crossing-over occurred at the four-strand stage, these flies were not truly trisomic; that is, the Y chromosome did not have the alleles for the genes on the X chromosomes. Rhoades (1933) used maize plants trisomic for chromosome 5 because this chromosome had a number of linked mutant genes and the $2n$ and $2n + 1$ plants could be readily identified by their different morphology to demonstrate crossing-over at the four-strand stage. Trisomic plants with a genotype of $+++/+++/bm\ pr\ v\text{-}2$ were crossed as female parent with a standard recessive (brown midrib, red seed, virescent seedling). Among the trisomic progenies, 4.1 percent were homozygous for *v-2*, 1.2 percent homozygous for *pr*, and none homozygous for *bm*. The trisomic homozygous for *v-2* or *pr* came from $n + 1$ eggs with a recessive allele in each chromosome 5. The recessive allele, however, was present in only one of the three chromosomes in the seed parent. In many of the $n + 1$ eggs with two *v-2* alleles, one chromosome had both the *bm* and the *pr* allele, indicating that one chromosome 5 was a recombinant and the other chromosome 5 a parental type. Such eggs could result only from a crossing-over between nonsister chromatids (Fig. 6-7). The *bm* locus on the short arm of this chromosome is so close to the centromere that a crossing-over between the locus and the centromere rarely occurs. The *pr* and *v-2* loci are on the other chromosome arm, and the *pr* locus is closer to the centromere than the *v-2* locus. Consequently, the trisomics homozygous for *pr* should be less frequent than the trisomics homozygous for *v-2*.

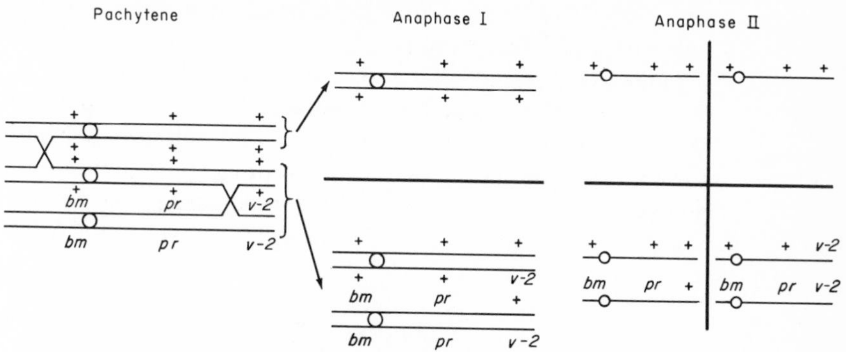

Fig. 6–7. Diagram illustrating the origin of mutant primary trisomic progeny from a cross between a simplex trisomic seed parent (+++/+++/*bm pr v-2*) and a mutant pollen parent (*bm pr v-2*/*bm pr v-2*) in maize. Chromatid crossing-over in the trisomic parent and only one kind of anaphase I disjunction are shown. The anaphase I segregation can yield one meiocyte with n and the other meiocyte with $n + 1$ chromosomes. At the anaphase II segregation, one $n + 1$ nucleus has two chromosomes 5 with the recessive *v-2* allele, but one chromosome is a crossover type and the other chromosome is a parental (noncrossover) type. (*After Rhoades, 1933.*)

The observed frequencies for trisomics homozygous for each mutant gene agreed with the relative genetic distance of each locus from the centromere.

Primary trisomics have been extensively used to assign mutant genes or linkage groups to specific chromosomes by noting the significant difference between the observed and expected frequencies for a monohybrid. Progeny from an autotriploid ($3n$) or from asynaptic or desynaptic mutants are screened for primary trisomics either by characteristically altered phenotypes or by chromosome count or both. The frequency of transmission for each trisome by the male or female gametophytes is established, and the number of trisomics in the progeny from self-pollinated trisomics is determined. The female gametophyte is generally more tolerant of an extra chromosome than the male gametophyte. These empirical parameters are used to calculate the expected frequency of recessive progeny from the simplex (*AAa*) trisomic parent previously obtained by crossing a homozygous dominant (*AAA*) trisomic seed parent and a recessive standard parent (Fig. 6-8). In practice, a number of mutant alleles is incorporated into one parent for crosses with the different dominant primary trisomics; the simplex (*AAa*) trisomic progeny are either self-pollinated or crossed with the recessive standard pollen parent; and the frequencies of recessive progeny are compared with

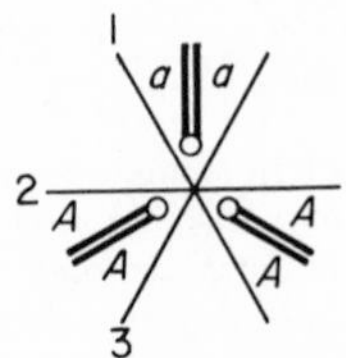

Plane	Metaphase I arrangement	First division products	Second division products
1	a a A A	A A / a a	A / a
	A A	A A	A
2	a a	a a	a
	A A A A	A A / A A	A / A
3	A A a a	A A / a a	A / a
	A A	A A	A

Fig. 6–8. Types of meiotic segregation for a trivalent in a heterozygous (*AAa*) primary trisomic. (*After Sturtevant and Beadle, 1939.*)

those calculated on the assumption that the locus is not in the trisome (Table 6-10). Although "simplex" has been used to indicate the number of dominant alleles, it seems reasonable to use this term to indicate the number of recessive alleles because the percentage of recessive progeny provides the critical evidence.

Secondary Trisomy

The extra chromosome in a secondary trisomic is an isochromosome. In the primary trisomics of *Datura stramonium,* the altered phenotypes represent the genetic imbalance caused by three alleles for each locus in the arms of the primary trisomes. The secondary trisomics, however, have four alleles for each locus in one arm and two alleles for loci in the other arm of the corresponding standard chromosome. For example, the echinus seed capsule and dimorphic pollen grains are pheno-

TABLE 6-10

Segregating Progeny from Crosses Involving Simplex or Duplex Trisomic Seed or Pollen Parents in Maize

Parents		Progeny		
			Recessive	
Seed	Pollen	Dominant	No.	%
Rr	*Rr*	608	204	25.3
Rr	*rr*	1,161	1,196	50.7
rr	*Rr*	132	135	50.6
rr	*Rrr*	1,392	2,685	65.9
rr	*RRr*	941	486	34.1
Rr	*RRr*	2,102	435	17.1
Rr	*Rrr*	348	182	34.4
Rrr	*Rr*	160	57	26.3
Rrr	*rr*	679	836	55.2
RRr	*rr*	819	213	20.6
RRr	*RRr*	584	65	10.0

Source: After McClintock and Hill, 1931.

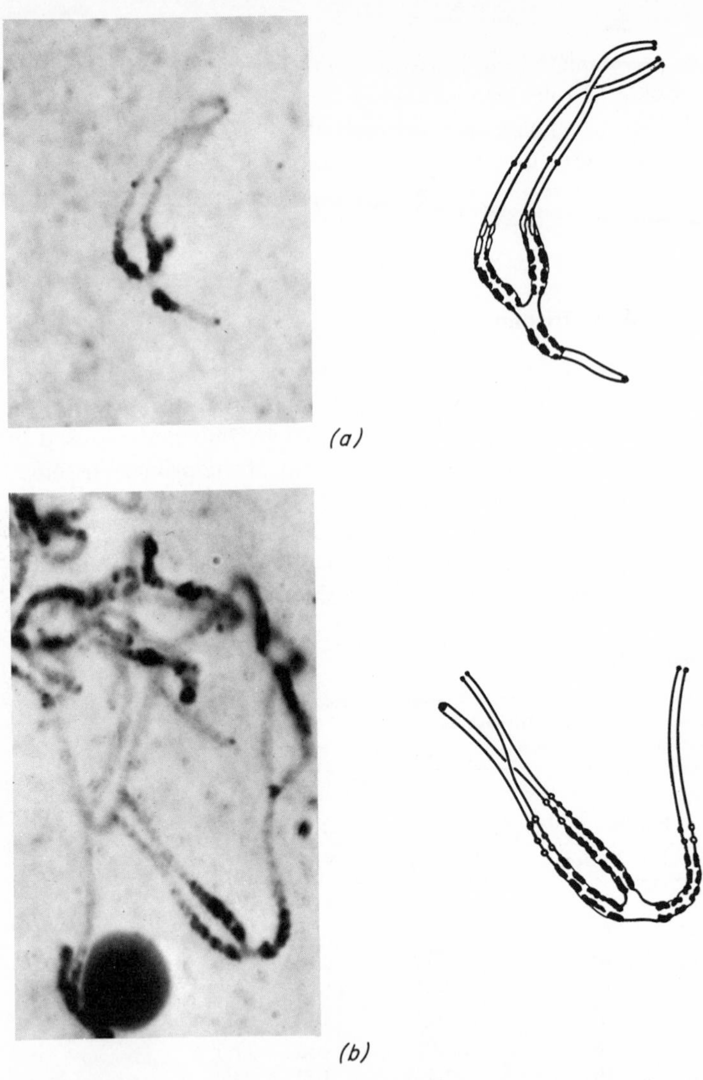

Fig. 6–9. Secondary trisome at pachytene in the tomato (*a*) Trivalent association in $2n$ + 8L·8L, showing one chiasma in the 8L arms. (*b*) Trivalent association in $2n$ + 12L·12L, showing one chaisma in the 12L arms. (*Photographs courtesy of Dr. C. M. Rick; Khush and Rick, 1969.*)

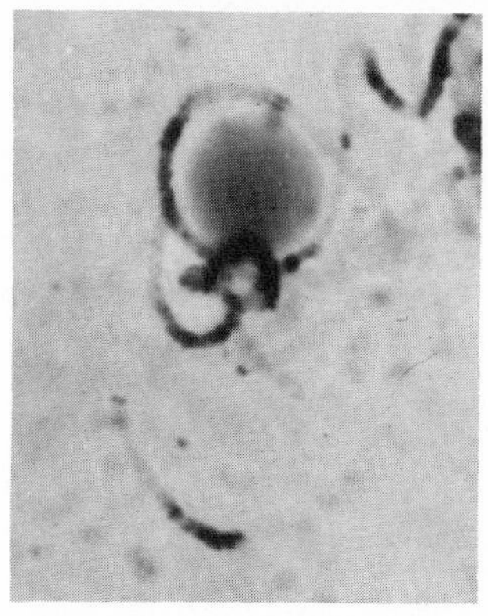

Fig 6–10. Univalent secondary trisome (isochromosome) 7S·7S at pachytene in the tomato, showing the terminally situated centromere. (*Photographs courtesy of Dr. C. M. Rick; Khush and Rick, 1967a.*)

types associated with primary trisome 5 (5·5′). The secondary trisomic 5·5 but not the complementary secondary 5′·5′ has only the dimorphic pollen grains, indicating that this phenotype is caused by a gene or genes in the ·5 arm of chromosome 5.

Khush and Rick (1969) have used secondary trisomes in the tomato (Figs. 6-9 and 6-10) for a monosomic analysis in a diploid species. This ingenious method for assigning genes to one or the other arm of specific chromosomes will be discussed in the section on compensating trisomy.

Tertiary Trisomy

The extra chromosome in a tertiary trisomic is a translocation chromosome from the 3:1 distribution of chromosomes in an interchange complex at anaphase I to give $n + 1$ gametophytes. Although tertiary trisomics have occasionally been used to place mutant genes in one or the other segment of the translocation chromosome, Khush and Rick (1967*b*) have been able to use tertiary trisomes (Fig. 6-11) in the tomato with considerable success. In this species, tertiary trisomics

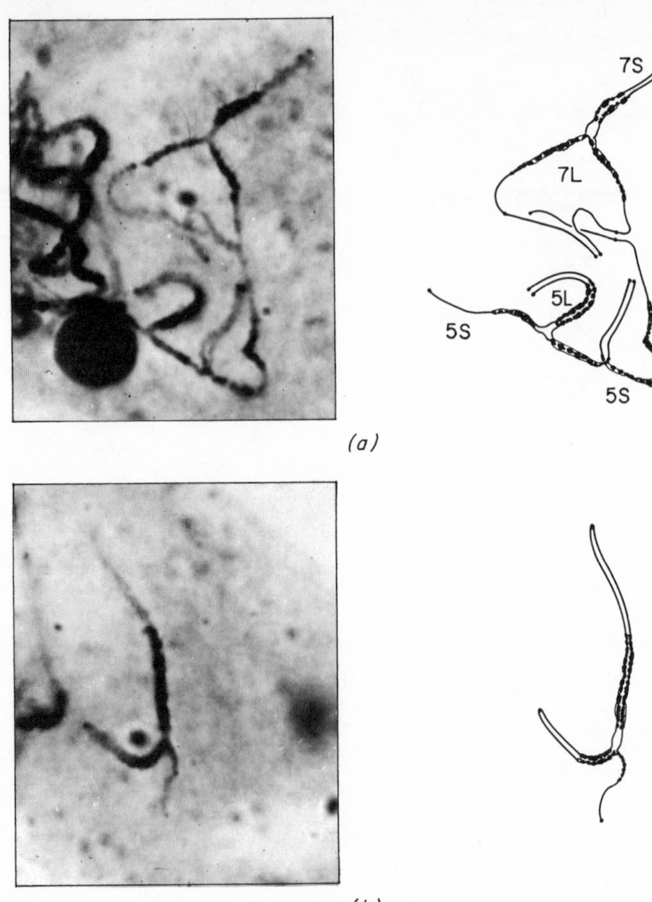

Fig. 6–11. (*a*) Tomato pachytene chromosomes showing an association of five chromosomes in a tertiary trisomic with $2n + 5S{\cdot}7L$. (*Photograph courtesy of Dr. C. M. Rick; Khush and Rick, 1967b.*) (*b*) Telocentric trisome for the short arm of chromosome 10 at pachytene. The centromeres of the three chromosomes are paired to form a unit. (*Photograph courtesy of Dr. C. M. Rick; Khush and Rick, 1968a.*)

are detected by their characteristically altered phenotypes, and the segments in the translocation chromosome are readily related to specific chromosomes at pachytene. By selecting translocation chromosomes with different arms rather than different segments, mutant genes were assigned to one or the other arm by comparing the calculated and observed frequencies of recessive standard or trisomic progeny from heterozygous trisomics which had been self-pollinated or crossed with a recessive standard pollen parent (Table 6-11).

TELOCENTRIC TRISOMY

Standard chromosomes can yield telocentric chromosomes with an intact or almost intact terminal centromere and one arm. Although these unusual chromosomes have been reported in several species, they are not usually exploited to place mutant genes in one arm of a chromosome. Rhoades (1936) and Khush and Rick (1968) have successfully used telocentric trisomy in maize and tomato (Fig. 6-11), respectively, for this purpose because the aneuploid plants are detected by their altered phenotype and the telotrisome can be identified as one or the other arm of a standard chromosome. For example, a heterozygous telocentric trisomic was synthesized in tomato to have the following genotype and chromosomal constitution (Table 6-12): 3L·3S *A* + 3L·3S *a* + ·3L *A*. Gametophytes with only the telocentric chromosome aborted, and those with a standard chromosome or with a standard and a telocentric chromosome were functional. Consequently, all the trisomic progeny from the self-pollinated heterozygous telotrisomic had the dominant phenotype, and the standard progeny segregated to yield a monohybrid ratio. The occasional recessive trisomic progeny represented a crossing-over between the telotrisome and the standard chromosome with the recessive allele. When the marker gene is not in the telotrisome, both standard and trisomic progeny segregate to yield a monohybrid ratio.

TABLE 6-11

Segregation Ratios in Five F_2 Progeny of Five Tertiary Trisomics in Tomato

				Progeny						
				2n			2n + 1			
					Recessive			Recessive		
Trisomic	Gene	Chromosome	Total *	Normal	No.	%	Normal	No.	%	X^2, 3:1
2n + 1L·11L	*scf*	1	500	334	84	20.0	82	0	0.0	24.72**
	inv	1	500	340	78	18.6	81	1	1.2	22.25**
2n + 4L·10L	*clau*	4	496	263	95	26.2	94	44	31.8	3.58
	ra	4	496	272	86	24.0	138	0	0.0	46.00**
	di	4	496	300	58	16.2	136	2	1.4	40.81**
	ful	4	239	107	33	23.5	73	26	26.2	1.17
	w-4	4	239	107	33	23.5	99	0	0.0	33.00**
	ag	10	202	91	25	21.5	83	3	3.4	21.80**
2n + 5S·7L	*mc*	5	150	82	26	24.0	42	0	0.0	14.00**
	tf	5	150	87	21	19.4	41	1	2.3	11.45**
	wt	5	150	72	36	33.3	27	15	35.7	2.57
2n + 7S·11L	*gs*	7	151	67	22	24.2	61	1	1.6	18.08**
2n + 9S·12S	*ah*	9	192	88	34	27.8	51	19	27.1	1.72
	marm	9	192	91	31	25.4	54	16	22.8	1.72

* The primary trisomics for the chromosome carrying the markers under study were not included in the total.

Source: Khush and Rick, 1967*b*.

TABLE 6-12

Segregation Ratios in F_2 Progeny for Three Telotrisomics in Tomato

Telotrisomic	Gene	Total	Progeny						
			2*n*			2*n* + 1			
				Recessive			Recessive		
			Normal	No.	%	Normal	No.	%	X^2, 3:1
2*n* + ·3L	*r*	184	94	38	28.7	39	13	25.0	0.00
	rv	180	102	18	15.0	59	1	1.6	17.41**
2*n* + ·7L	*var*	176 *	77	31	28.7	52	15	22.3	0.22
2*n* + ·8L	*cpt*	131 †	42	16	27.5	65	1	1.5	19.39**

* One plant in the family was triplo-7.

† Nine plants in the family were 2*n* + ·8L + ·8L.

Source: Khush and Rick, 1968.

COMPENSATING TRISOMY

In one type of compensating trisomy, one standard chromosome has been broken and each segment is present in different translocation chromosomes. Furthermore, each translocation chromosome has a segment from different chromosomes. In such a compensating trisomic, the chromosome formula would be $2n - 1{\cdot}1' + 1{\cdot}2 + 1'{\cdot}3$ and the chromosome number, $2n - 1$.

A trisomic plant of *D. stramonium* with the nubbin seed-capsule phenotype was found among the plants grown from seed exposed to radium (Blakeslee, 1927). The breeding data and the chromosome associations at diakinesis and metaphase I for this mutant indicated that this trisomic was not primary, secondary, or tertiary. In the nubbin trisomic, one standard chromosome is represented by segments in two different translocation chromosomes, and the other segments of the translocation chromosomes are duplications. The cytogenetic analysis of the nubbin phenotype illustrates an ingenious integration of morphological, breeding, and chromosomal observations to determine the cytogenetic formula of a compensating trisomic.

The cross between nubbin and a standard pollen parent gave mostly standard and nubbin progeny and the following primary or tertiary trisomics (Fig. 6-12): buckling (3·3′), echinus (5·5′), pinched (1′·3), or hedge (1·5). Crosses between the tertiary trisomics (pinched, hedge) and a standard pollen parent yielded standard progeny and different trisomics, particularly the rolled primary trisomic (1·1′). The absence of the rolled trisomic in the progeny from nubbin was a significant clue in understanding the chromosomal constitution of nubbin.

The morphological characteristics of nubbin were compared with those of the pertinent primary and secondary trisomics. Nubbin had the dimorphic pollen grains found in the mutilated secondary trisomic (5·5) and certain characteristics of the strawberry secondary trisomic (3·3) but not of the complementary aerolate secondary trisomic (3′·3′). While these ob-

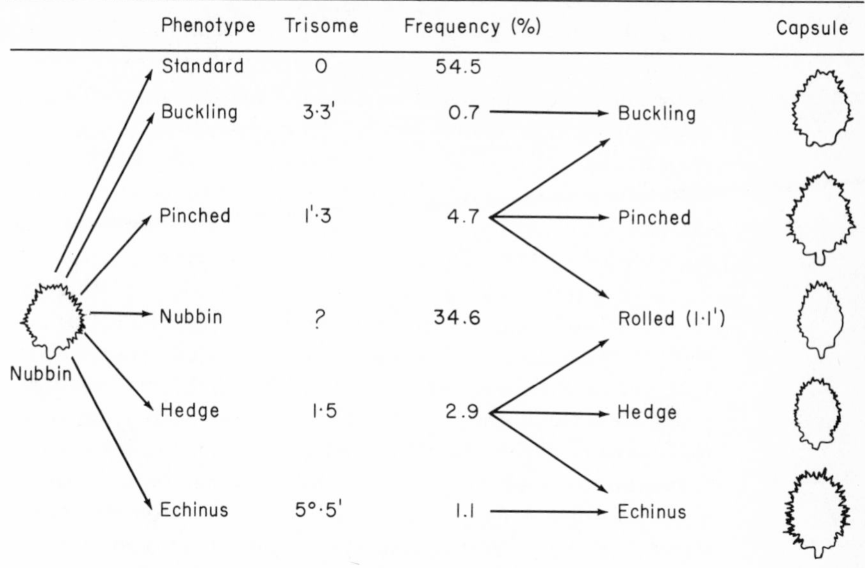

Fig. 6–12. Inheritance of different types of progeny and their frequencies from the cross between a nubbin mutant seed parent and a standard pollen parent in *Datura stramonium.* Note the absence of the rolled trisomic in progeny from nubbin and its presence in progeny from the pinched and hedge trisomics. (*A. F. Blakeslee. 1927. Annals of the New York Academy of Sciences, vol. 30,* fig. 2, p. 4.)

servations suggested that nubbin might have been a tertiary trisomic (3·5), a chain of seven chromosomes was found at metaphase I. To account for all the breeding data and the chromosome association, nubbin was assigned the following chromosomal formula: 5·5′ + 5′·5 + 5·1 + 1·1′ + 1′·3 + 3·3′ + 3′·3. The haploid gametophytes have the standard chromosomes 5·5′ + 1·1′ + 3·3′ and the $n + 1$ gametophytes two standard chromosomes (5·5′ + 3·3′) and two translocation chromosomes (5·1 + 1′·3) to ensure the equivalence of a standard chromosome 1 (1·1′). The distribution of the chromosomes of the interchange complex of seven chromosomes can be manipulated to give low frequencies of the primary (buckling, echinus) and tertiary (pinched, hedge) trisomics but not of the rolled primary trisomic (1·1′). The distribution of chromosomes from the association of five chromosomes in the tertiary trisomics, however, can produce $n + 1$ gametophytes with an extra standard chromosome 1 (1·1′).

The cytogenetic analysis of the nubbin compensating trisomic in *D. stramonium* did not consider Mendelian genes or the precise length of the segments in the translocation chromosomes. Khush and Rick (1967*a*) synthesized novel compensating trisomics in the tomato, which were assigned cytogenetic formulas to indicate the precise length and source of the extra chromosomal segments. For example, chromosome 3 had a short (S) and a long (L) arm, and the isochromosomes or telocentric chromosome from this standard chromosome involved an entire chromosome arm. Two types of compensating trisomics were obtained: $2n$ — 3S·3L + 3L·3L + ·3S and $2n$ — 3S·3L + 3S·3S + 3L·3L. One standard chromosome was replaced, in the first type, by an isochromosome and a telocentric chromosome, and in the second type, by two complementary isochromosomes. The characteristic morphological alterations associated with specific trisomics in this species were responsible for the detection of unusual compensating trisomics, and their chromosomal constitution was established by examining the pachytene chromosomes and associations at diakinesis (Fig. 6-13).

The isochromosome + telocentric compensating trisomics were used to place mutant genes in specific chromosome arms by essentially a monosomic analysis (Table 6-13). By crossing a heterozygous trisomic as seed parent with a recessive standard parent, the diploid progeny exhibited the recessive phenotype and the compensating trisomic progeny exhibited the dominant phenotype. Furthermore, the instability of the telocentric chromosome in somatic nuclear divisions gave a mosaic of somatic tissues in which the cells with the telocentric chromosome with

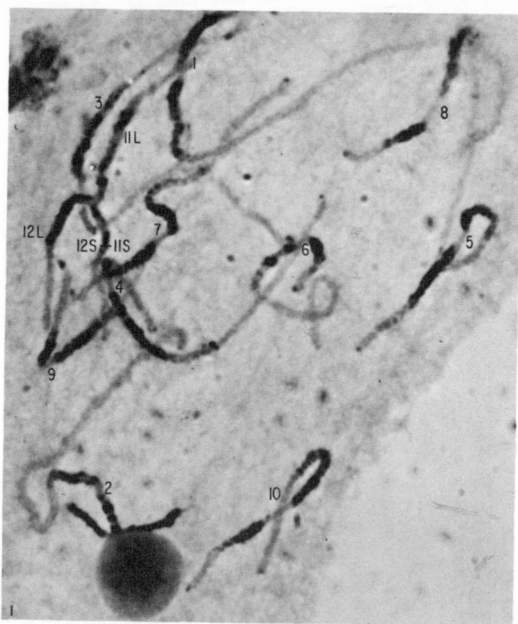

Fig. 6–13. The chromosome complement of tomato at pachytene for the tertiary monosomic haplo-11S·12S, showing a trivalent formed by chromosomes 11, 12, and 11L·12L as well as normal pairs for the other chromosomes. Each chromosome is numbered at its centromere except for the associations of chromosomes 11 and 12. The chromosome arms 11L and 12L are paired normally, but the single 11S and 12S arms are paired nonhomologously for their full lengths. (*Photograph courtesy of Dr. C. M. Rick; C. M. Rick and G. S. Khush. 1968. Cytogenetic explorations in the tomato genome, pp. 45–68, fig. 1, p. 46. In R. Bogart (ed.), Seminars in genetics, vol. 1. Oregon State University Press, Corvallis.*

the dominant allele exhibit the dominant phenotype, and cells without this chromosome exhibit the recessive phenotype. This method works efficiently for phenotypes which can be scored in vegetative tissue.

TABLE 6-13

Progeny of the Testcross between a Compensating Trisomic for Chromosome 3 and the Diploid Marker Stock in Tomato

Constitution for Chromosome 3:
[*3S·3L* (ru) + *3L·3L* + *·3S* (ru+)] × [*3S·3L* (ru) + *3S·3L* (ru)]

Expected Type	Constitution for Chromosome 3	Phenotype	Observed No.	Observed %
Diploid	3S·3L (*ru*) + 3S·3L (*ru*)	*ru*	103	56.3
Compensating trisomic (parental type)	3S·3L (*ru*) + 3L·3L (*ru*) + ·3S (*ru*+)	+	74	40.5
Telotrisomic	3S·3L (*ru*) + 3S·3L + ·3S (*ru*+)	+	5	2.7
Secondary trisomic	3S·3L (*ru*) + 3S·3L (*ru*) + 3L·3L	*ru*	0	0
Compensating trisomic (double isotrisomic)	3S·3L (*ru*) + 3S·3S (*ru*+/*ru*+) + 3L·3L	+	1	0.5

Source: Khush and Rick, 1967*a*.

The compensating trisomic with complementary isochromosomes provides an interesting means of substituting one chromosome from one species of *Lycopersicon* for one chromosome in *L. esculentum* (Fig. 6-14). Although such a substitution has been accomplished by nullisomics in *T. aestivum*, a hexaploid species, no method was available for a diploid species. The scheme assumes that the substituted chromosome from the donor species will not be involved in a crossing-over with the isochromosomes from *L. esculentum*, and the available evidence suggests that the frequency of such a crossing-over should be minimal.

Although primary trisomy is the most frequently encountered aneuploid, other types of trisomy can be expected to occur in the course of an extensive and intensive cytogenetic study. Their unique characteristics can be exploited to place mutant genes in specific regions of a chromosome whenever

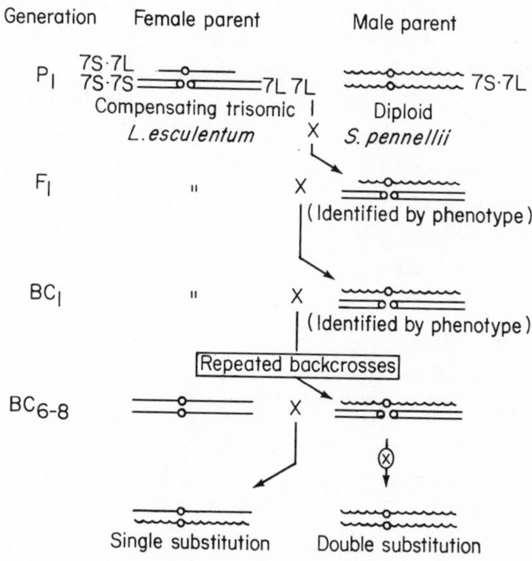

Fig. 6–14. Scheme for substituting chromosome 7 of the wild species *Solanum pennellii* in the complement of the tomato *Lycopersicon esculentum*, using a compensating trisomic. (*Khush and Rick, 1967a.*)

the species has good to excellent pachytene chromosomes, reasonably comprehensive linkage maps, and a genotype sufficiently sensitive to the chromosomal imbalance so that each trisomic type will be characterized by specific phenotypic alterations. The detailed cytogenetic analysis of the tomato has provided model situations for similar studies in other species with the same array of desirable characteristics.

TETRASOMY

The self-pollination of a primary trisomic in a diploid or tetraploid species usually yields diploid, trisomic, and rarely tetrasomic ($2n + 1^{II}$) progeny. A chromosome count of progeny from a self-pollinated autotriploid frequently reveals individuals with two extra chromosomes, but they are usually double primary trisomics ($2n + 1 + 1$). The genetic imbalance caused by the extra bivalent is probably responsible for the improper development of tetrasomic zygotes or embryos and for their absence in progeny from self-pollinated primary trisomics of diploid or tetraploid species but not of hexaploid species (Table 6-14). Tetrasomics were used to compensate for certain nullisomes for demonstrating homoeologous chromosomes in the hexaploid species *T. aestivum*. Tetrasomic inheritance has been investigated in autotetraploids ($4n$) which are genetically balanced.

TABLE 6-14

Frequencies of Tetrasomics in Progenies from Four Self-pollinated Trisomics of *Triticum aestivum*

Chromosome	Total	Cytologically Analyzed	Tetrasomic	
			No.	%
IV	78	73	3	3.8
VIII	130	115	2	1.5
XVI	183	49	1	0.5
XVIII	109	26	2	1.8

Source: Sears, 1954.

REFERENCES

AVERY, A. G., S. SATINA, AND J. RIETSEMA. 1959. Blakeslee: the genus *Datura*. Ronald Press, New York.

BLAKESLEE, A. F. 1921. The Globe mutant in the jimson weed (*Datura stramonium*). Genetics 6:241–264.

———. 1927. Nubbin: a compound chromosomal type in *Datura*. Ann. N.Y. Acad. Sci. 30:1–29.

BRIDGES, C. B. 1916. Nondisjunction as proof of the chromosomal theory of heredity. Genetics 1:1–52, 107–163.

BUCHHOLZ, J. T., AND A. F. BLAKESLEE. 1930. Pollen-tube growth and control of gametophyte selection in cocklebur, a 25-chromosome Datura. Bot. Gaz. 90:366–383.

CLAUSEN, R. E., AND D. R. CAMERON. 1944. Inheritance in *Nicotiana tabacum*. XVIII. Monosomic analysis. Genetics 29:447–477.

KHUSH, G. S., AND C. M. RICK. 1967*a*. Novel compensating trisomics of the tomato: cytogenetics, monosomic analysis, and other applications. Genetics 55:297–307.

——— AND ———. 1967*b*. Tomato tertiary trisomics: origin, identification, morphology and use in determining position of centromeres and arm location of markers. Can. J. Genet. Cytol. 9:610–631.

——— AND ———. 1968. Tomato telotrisomics: origin, identification, and use in linkage mapping. Cytologia 33:137–148.

——— AND ———. 1969. Tomato secondary trisomics: origin, identification and use in cytogenetic analysis of the genome. Heredity 24:129–146.

MCCLINTOCK, B., AND H. E. HILL. 1931. The cytological identification of the chromosome associated with the R-G linkage group in *Zea mays*. Genetics 16:175–190.

MORRIS, R., AND E. R. SEARS. 1967. The cytogenetics of wheat and its relatives, pp. 19–87. *In* K. S. Quisenberry and L. P. Reitz (eds.), Wheat and wheat improvement. American Society of Agronomy, Madison.

OLMO, H. P. 1935. Genetical studies of monosomic types of *Nicotiana tabacum*. Genetics 20:286–300.

RHOADES, M. M. 1933. An experimental and theoretical study of chromatid crossing over. Genetics 18:535–555.

———. 1936. A cytogenetic study of a chromosome fragment in maize. Genetics 21:491–502.

———. 1955. The cytogenetics of maize, pp. 123–219. *In* G. F. Sprague (ed.), Corn and corn improvement. Academic, New York.

Rick, C. M., and D. W. Barton. 1954. Cytological and genetical identification of the primary trisomics of the tomato. Genetics 39:640–666.

——— and G. S. Khush. 1968. Cytogenetic explorations in the tomato genome, pp. 45–68. *In* R. Bogart (ed.), Seminars in genetics. Oregon State University Press, Corvallis.

Sears, E. R. 1953. Nullisomic analysis in common wheat. Amer. Natur. 87:245–252.

———. 1954. The aneuploids of common wheat. Missouri Agr. Exp. Sta. Res. Bull. 572.

Sturtevant, A. H., and G. W. Beadle. 1939. An introduction to genetics. Saunders, Philadelphia.

SUPPLEMENTARY REFERENCES

Blakeslee, A. F., and A. G. Avery. 1938. Fifteen-year breeding records of 2N + 1 types in *Datura stramonium*, pp. 315–351. *In* Cooperation in research. Carnegie Inst. Washington Pub. 501.

Bridges, C. B. 1921. Genetical and cytological proof of non-disjunction of the fourth chromosome of *Drosophila melanogaster*. Proc. Nat. Acad. Sci. 7:186–192.

Kuspira, J., and J. Unrau. 1959. Theoretical ratios and tables to facilitate genetic studies with aneuploids. Can. J. Genet. Cytol. 1:267–312.

Sears, E. R. 1965. Nullisomic-tetrasomic combinations in hexaploid species. *In* R. Riley and K. R. Lewis (eds.), Chromosome manipulations and plant genetics. Heredity (Suppl.) 20:29–45.

———. 1969. Wheat cytogenetics. Ann. Rev. Genet. 3:451–468.

7 EUPLOIDY

Many plant genera include species with different chromosome numbers which are frequently multiples of a basic number. For example, the genus *Triticum* has species with gametophyte chromosome numbers of 7, 14, or 21. In such genera, a species with two genomes is a basic diploid (2X), one with four genomes a tetraploid species (4X), one with six genomes a hexaploid species (6X), and so forth. For *Triticum,* the sporophyte of a diploid species has 7 bivalents, the tetraploid 14 $(7 + 7)$ bivalents, and the hexaploid 21 $(7 + 7 + 7)$ bivalents.

In basic diploid species, the gametophyte has one genome, and the sporophyte has two homologous genomes. In lower plants, the gametophytes and sporophytes are readily distinguishable, but in higher plants, the female gametophyte is inconspicuous as the embryo sac and the male gametophyte as a pollen grain while the sporophyte is the conspicuous form. Sporophytes with one, three, or four genomes have been found in basic diploid species, and the terminology for these considers both the number and homology of the genomes. Sporophytes with one genome are *monoploids* rather than haploids; sporophytes of polyploid species with the gametophytic chromosome number are termed *polyhaploids* rather than monoploids. Sporophytes with three homologous genomes are *autotriploids* (3*n*), and those with four homologous genomes *autotetraploids* (4*n*).

MONOPLOIDY

Monoploids regularly occur as individuals in the bryophytes and in hymenopteran insects. In higher plants, monoploid sporophytes are relatively uncommon and generally differ from the diploids by their reduced stature, slender proportions, smaller organs, and almost complete sterility. Attempts to detect monoploids in the field are hindered by their reduced vigor and inability to compete with the diploids. In *Datura stramonium,* the frequency of spontaneous monoploids has been estimated to be approximately 5×10^{-3}; and in maize, approximately 2×10^{-3}. An extensive study in maize indicated that the frequency of monoploids in this species is genetically determined (Chase, 1969). A deliberate search for monoploids can be accomplished in hybrid progeny by detecting recessive seedlings with the maternal genotype determined by mutant alleles in different chromosomes. Suspect monoploids are confirmed by a chromosome count.

Several techniques have been developed to enhance the frequency or detection of monoploids. In amphibians, the egg can be stimulated by physical or chemical agents to initiate mitosis, thereby simulating the stimulus provided by fertilization. A similar method has been used for plants by using pollen incapable of effecting a fertilization such as pollen from a related species, inactivated by high dosages of x-ray or stored prior to pollination. In the genus *Crepis,* hybrid seedlings from the cross between *C. tectorum* and *C. capillaris* die, but the monoploid seedlings with the maternal genome survive. Seeds with two embryos have been found in many species, and one seedling from such seeds may be monoploid with the maternal genome.

Although monoploids are generally sterile, spontaneous chromosome doubling can occur during embryogenesis or the development of the sporophyte. A sample of 110 root tips from monoploids of *C. capillaris* included 28 tips with at least one diploid cell and 42 tips with only diploid cells. Consequently, monoploids may yield seeds after self-pollination when diploid cells are in the sporogenous tissue. Methods for inducing chromosome doubling can be used to increase seed production.

Diploid progeny from monoploid plants represent fertile plants which are homozygous for all loci. In maize, extensive inbreeding is necessary to select lines for the production of hybrid seed. Monoploids have provided a shortcut in the program for finding homozygous diploid lines, which are then tested for their value in the production of hybrid seeds.

The chromosomes in the monoploids are more or less randomly distributed at anaphase I, and both the microspores and megaspores are deficient. Approximately 12 percent of the pollen grains from a monoploid in *D. stramonium,* however, were normal in appearance, and a cytological examination indicated approximately the same value for dyads at the end of microsporogenesis. The dyads were produced by microsporocytes which directly entered the second meiotic division and gave functional haploid spores.

One or more bivalents at diakinesis or metaphase I have been reported in monoploids. In these cases, each bivalent had at least one chiasma. Polyhaploids may have also bivalents at these stages. The bivalents in monoploids or in polyhaploids indicate homologous segments and furnish visual evidence of duplications in different chromosomes.

AUTOTRIPLOIDY

Spontaneous autotriploids are occasionally found in the progeny of diploid species, presumably as the products of the fertilization of haploid and diploid gametes or gametic nuclei. Autotriploids are usually obtained by crossing an autotetraploid, generally as the female parent, and a diploid. In some species, the cross is difficult or often fails.

The reduced fertility of autotriploids is directly related to the distribution of the chromosomes during meiosis. Each trio of homologous chromosomes may form a trivalent, a bivalent and univalent, or three univalents, depending on the number of chiasmata for each chromosome. Longer chromosomes are more likely than shorter chromosomes to form trivalents. The distribution of the chromosomes in a trivalent or of the univalents at anaphase I or II is responsible for the high frequency of aneuploid gametophytes. The most comprehensive study of chromosome distribution at each meiotic division during microsporogenesis and megasporogenesis and of the chromosome numbers for the gametophytes was accomplished for autotriploids in *D. stramonium* (Satina and Blakeslee, 1937*a,b;* Satina, Blakeslee, and Avery, 1938).

Most of the gametophytes produced by an autotriploid are aneuploid. While the range of chromosome numbers in the functional gametophytes is dependent on the genetic constitution of a species, the haploid, $n + 1$, $n + 1 + 1$, and $2n$ gametophytes are usually viable and functional. The fertility of autotriploids is also related to the genetic imbalance which can be tolerated by the aneuploid zygotes or developing embryos. Consequently, the range of chromosome numbers in the progeny from a self-pollinated autotriploid or from crosses with the standard plants as seed or pollen parent depends on the species.

The number of chiasmata at diakinesis or metaphase I can be greater for the autotriploid than for the corresponding diploid. Chromosomes with two or more chiasmata in one arm often have fewer chiasmata in different meiocytes because interference reduces the potential number of chiasmata. While pairing involves only two homologous segments, the homologous chromosomes in a trivalent may switch partners to pro-

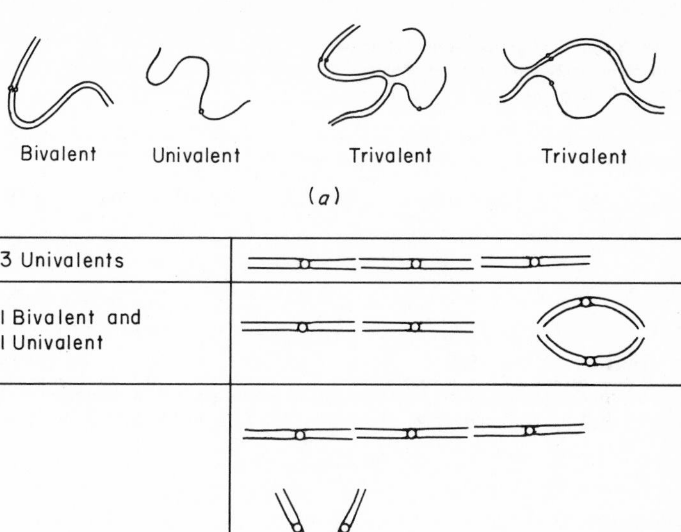

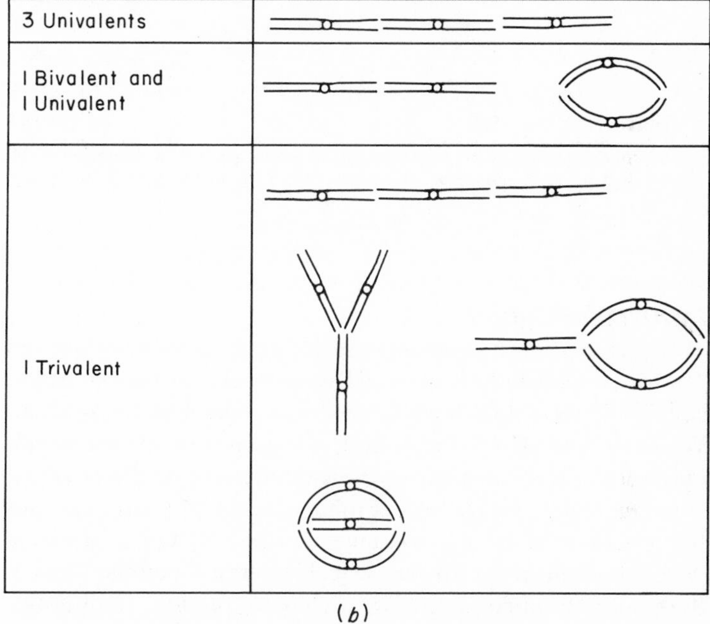

Fig. 7–1. (*a*) Pairing relations for three homologous chromosomes at pachytene. (*b*) Configurations for three homologous chromosomes at diakinesis.

duce different types of trivalent associations (Fig. 7-1). Several methods have been used to calculate the number of chiasmata at diakinesis or metaphase I for autotriploids to obtain values to compare with those for the corresponding diploid (Table 7-1).

TABLE 7-1

Chromosome Configurations and Number of Chiasmata at Diakinesis in 50 Pollen Mother Cells of an Autotriploid Species ($3n = 36$) *

Number of	Chromosome Associations			
Chiasmata	Univalents	Bivalents	Trivalents	Total
0	356	...	...	0
1	...	155	...	155
2	...	195	221	832
3	...	...	27	81
Total	356	350	248	1,068

* Values for comparative purposes: 21.4 chiasmata per pollen mother cell or 0.59 chiasma per chromosome or 1.35 chiasmata per chromosome in bivalents and trivalents.

AUTOTETRAPLOIDY

The occasional failure of a cell to enter anaphase is responsible for a tetraploid nucleus with four homologous genomes. When this event happens during embryogenesis or the development of the sporophyte, a diploid-autotetraploid mosaic, or chimera, results. The incorporation of autotetraploid cells into the germinal tissue of seedlings or the flowering branches yields autotetraploid flowers, which after self-pollination produce autotetraploid seeds. The occasional production of a diploid gametophyte by a nonreduction during meiosis is more likely to give an autotriploid, which in turn after self-pollination often yields autotetraploids among the progeny.

The nuclei and cells of autotetraploids generally are larger than those of the corresponding diploids, and the greater volume of the cells in an autotetraploid can be reflected in the dimensions of the polyploid or its organs. In plants, the autotetraploid generally develops more slowly and is more robust and larger than the diploid parent. Furthermore, the autotetraploid has thicker stalks, larger and deeper green leaves, bigger stomata and guard cells, larger flowers, and obviously bigger pollen grains. The usual method of detecting autotetraploids involves measurements of the length of guard cells or stomata or the diameter of pollen grains to compare with similar measurements for the corresponding diploid. The final decision, however, rests on the chromosome count. In comparing diploid, autotriploid, and autotetraploid plants of one species, the increase in stature and in the size of certain organs of the autotriploid is usually intermediate between the diploid and autotetraploid. The *semigigas* and *gigas* mutants in *Oenothera lamarckiana* found by de Vries were an autotriploid and an autotetraploid, respectively.

Cytogeneticists and plant breeders have been interested in the induction of autopolyploids for several reasons. Autotriploids yield primary trisomics which are useful in placing mutant genes in specific chromosomes. Interspecific hybridizations between a diploid and a tetraploid species may fail to produce seed, whereas the hybridization between the autotetraploid and the tetraploid species sometimes succeeds in producing viable seeds. After chromosome doubling, sterile interspecific hybrids usually produce viable seeds. Consequently, the induction of polyploidy is an area of considerable interest and effort. In the Solanaceae (tobacco, tomato, and *D. stramonium*), a callus is produced at the cut surface when the main stem of vigorously growing plants is decapitated. A cytological survey of shoots emerging from the callus usually reveals autotetraploids, and the self-pollinated flowers from such shoots yield autotetraploid seeds. When temperature shocks are applied to developing embryos or seedlings, the frequency of somatic autotetraploid cells is enhanced so that diploid-autotetraploid chimeras are likely to occur. The most effective agent for doubling the chromosome number of somatic cells is the alkaloid compound colchicine applied to germinating seeds, seedlings, or branches by numerous methods (Eigsti and Dustin, 1955). Although other chemicals have been found to induce chromosome doubling in plants, colchicine remains the most effective compound.

The cytogenetics of autotetraploidy has been extensively studied in plants because autotetraploids are easily obtained, vigorous, fertile, and amenable to cytological analysis. The relatively few autotetroploids reported for animal species develop poorly or are sterile.

Four homologous chromosomes can form a quadrivalent, a trivalent and univalent, two bivalents, or a bivalent and two univalents at diakinesis and metaphase I. The relative frequencies of these chromosome associations in the meiocytes, usually pollen mother cells, at metaphase I are determined by such factors as chromosome length, arm ratio, and the rate of terminalization. These factors also determine the configurations for the quadrivalents (Fig. 7-2). Consequently, the frequencies vary from meiocyte to meiocyte from one autotetraploid plant and from species to species. For example, quadrivalents are less frequent than bivalents in autotetraploid plants of tomato, while quadrivalents and bivalents are common in autotetraploid plants of maize.

The distribution of the chromosomes in each meiotic division for an autotetraploid with quadrivalents, trivalents, bivalents, and univalents is responsible for spores with chromosome numbers ranging from less to more than the diploid number. The functional gametophytes produce zygotes and embryos with chromosome numbers different from the autotetraploid number. In a population of 54 progeny from a self-pollinated autotetraploid in maize, the chromosome numbers ranged from 37 to 42: 37 (1), 38 (3), 39 (6), 40 (27), 41 (12), and 42 (5). The genetic imbalance in plants with less or more than the autotetraploid chromosome number is tolerated to different degrees by different species. Within one species, autotetraploids are likely to be less fertile than the diploid but more fertile than the autotriploid. An increase in the number of bivalents at the expense of quadrivalents and a reduction in the number of trivalents and univalents enhance the fertility of autotetraploids.

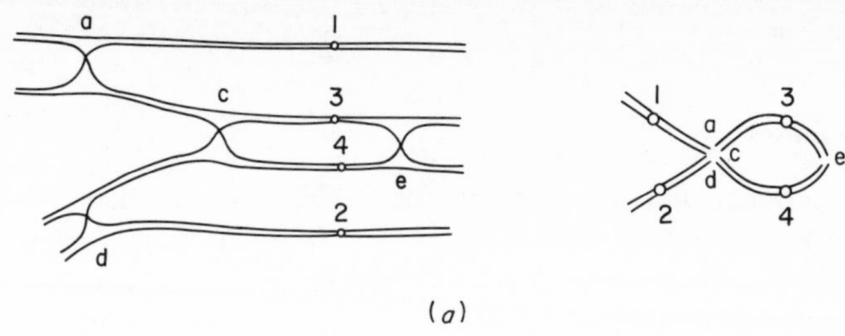

(a)

4 Univalents	
2 Bivalents	
1 Univalent 1 Bivalent	
1 Trivalent 1 Univalent	
1 Quadrivalent	

(b)

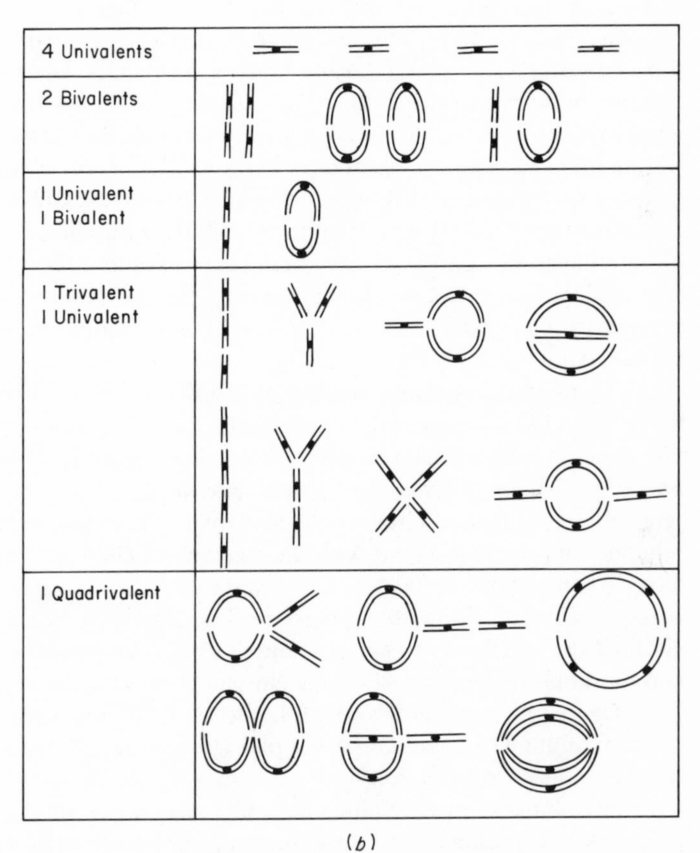

Fig. 7–2. (*a*) One type of chromosome association (quadrivalent) for four homologous chromosomes at diplotene, showing the sites of four chiasmata, and after terminalization, at diakinesis. (*b*) Configurations for four homologous chromosomes at diakinesis.

Genetics

The range of heterozygosity for a locus is extended for the autotetraploid. The terminology for heterozygosity adopted for trisomy has been applied to autotetraploids in which each chromosome is tetrasomic. The diploid has one type of heterozygote (*Aa*), and the trisomic has two types: simplex (*AAa*) or duplex (*Aaa*). In an autotetraploid, a heterozygote can be simplex (*AAAa*), duplex (*AAaa*), or triplex (*Aaaa*). The fertility of autotetraploids of different numbers of recessive alleles is responsible for our knowledge of tetrasomic inheritance. Burnham (1962) has presented a thorough discussion of the genetics of autotetraploids. The calculated frequencies of the functional gametophytes from autotetraploids with each type of heterozygosity are based on a number of assumptions: (1) the production of diploid gametophytes, (2) chromosome or chromatid segregation, and (3) a single dominant allele yielding the dominant phenotype.

Chromosomal segregation for a simplex (*AAAa*) autotetraploid yields gametophytes with both dominant alleles or one dominant and one recessive allele. Consequently, such plants would not be expected to yield mutant progeny from a self-pollination or a testcross without invoking meiotic nondisjunction for the chromosome with the recessive allele. When the locus is sufficiently distal from the centromere, crossing-over is responsible for placing the recessive alleles, formerly in sister chromatids, on different chromosomes. The segregation of the chromosomes during meiosis can yield gametophytes with two homologous chromosomes and both recessive alleles. Consequently, mutant progeny from simplex autotetraploids result from chromatid segregation.

Early interest in the induction and cytogenetics of autotetraploids was stimulated by the potential value of autotetraploids in producing new varieties of ornamental, horticultural, or crop species. For example, autotetraploids of ornamental species usually have larger flowers than those of the corresponding diploid varieties. Certain phenotypic alterations are found in autotetraploids such as increased vegetative growth, changed chemical composition, delayed maturity—to mention a few—which can be exploited whenever they have an economic value. Unfortunately, the promise of the early investigations on this aspect of autotetraploidy has not been fully realized.

REFERENCES

BURNHAM, C. R. 1962. Discussions in cytogenetics. Burgess, Minneapolis.

CHASE, S. S. 1969. Monoploids and monoploid-derivatives of maize (*Zea mays* L.). Bot. Rev. 35:117–167.

EIGSTI, O. J., AND A. P. DUSTIN. 1955. Colchicine in agriculture, medicine, biology, and chemistry. Iowa State College Press, Ames.

SATINA, S., AND A. F. BLAKESLEE. 1937*a*. Chromosome behavior in triploid *Datura stramonium*. I. The male gametophyte. Amer. J. Bot. 24:518–527.

——— AND ———. 1937*b*. Chromosome behavior in triploid *Datura stramonium*. II. The female gametophyte. Amer. J. Bot. 24:621–627.

———, ———, AND A. G. AVERY. 1938. Chromosome behavior in triploid *Datura stramonium*. III. The seed. Amer. J. Bot. 25:595–602.

SUPPLEMENTARY REFERENCES

BELLING, J., AND A. F. BLAKESLEE. 1923. The reduction division in haploid, diploid and tetraploid Daturas. Proc. Nat. Acad. Sci. 9:106–111.

BLAKESLEE, A. F., J. BELLING, AND M. E. FARNHAM. 1923. Inheritance in tetraploid Daturas. Bot. Gaz. 76:329–373.

Dawson, G. W. P. 1962. An introduction to the cytogenetics of polyploids. Davis, Philadelphia.

Gilles, A., and L. F. Randolph. 1951. Reduction of quadrivalent frequency in autotetraploid maize during a period of 10 years. Amer. J. Bot. 38:12–17.

Kimber, G., and R. Riley. 1963. Haploid angiosperms. Bot. Rev. 29:480–531.

Ramanujam, S., and N. Parthasarathy. 1953. Autopolyploidy. Indian J. Genet. Plant Breed. 13:53–82.

Randolph, L. F. 1935. Cytogenetics of tetraploid maize. J. Agr. Res. 50:591–605.

8 MAMMALIAN CYTOGENETICS

The choice of a species for cytogenetic studies considers such factors as a relatively low chromosome number, good to excellent pachytene chromosomes, comprehensive linkage groups, unique breeding, cytological or genetic characteristics, and inherent economic significance. For many years, mammals did not offer any advantages not already present in *Drosophila melanogaster* or certain plant species. Laboratory mammalian species had been used for genetic studies to provide mutants for biochemical, physiological, or anatomical studies which might be applied to a specific mammal, *Homo sapiens.*

Early attempts to determine chromosome numbers and morphology in mammals were restricted to sectioned somatic or germinal tissue. The varying counts for a species were attributed to sliced chromosomes in the serial sections. For example, a modal chromosome number of 48 was reported for man and, in time, this incorrect estimate was accepted as a fact. The development of mammalian cytogenetics can be attributed to advances in tissue culture and to the discovery of chromosome diseases in man. Somatic cells are grown in culture, treated with colchicine to increase the number of cells at metaphase, suspended in a hypotonic solution, spread in a single optical plane on a slide, and stained. This technique gives cells with intact chromosomes which can readily be counted and described. Furthermore, this method can be used to process cells from different issues. The association between a primary trisome and Down's syndrome (mongolism) in man stimulated a search for other chromosomal diseases. The currently rapid growth of mammalian cytogenetics was stimulated by medical interests. The relatively straightforward method of obtaining suitable material for counting and describing mammalian chromosomes has been a major factor in the accumulation of cytological data for numerous mammalian species.

Mammalian cytogenetics is currently restricted to relatively few species, notably man and mouse, but this situation will undoubtedly change in the near future. Mammals generally have relatively high chromosome numbers, and few metaphase chromosomes can be identified by their morphology. Opportunities to examine germinal cells for chromosomes at pachytene, diakinesis, or metaphase I are usually limited. The cytological survey of mammals, however, has uncovered a few species with a low number of distinguishable chromosomes which can be exploited for cytogenetic studies with either the living animal or cells in culture.

Although mouse cytogeneticists seem to have a distinct advantage in working with a laboratory species, human cytogeneticists have access to a greater supply of chromosome aberrations in a species now yielding numerous mutants, particularly for variant proteins or enzymes. Many human chromosomal aberrations produce diseases sufficiently severe to warrant institutional care. In some cases, the aberrations have been associated with previously described syndromes. For these reasons, contemporary mammalian cytogenetics has become more concerned with man than with other species.

The cytological survey of mammalian species has been vigorous but mostly descriptive. The chromosome numbers and karyotypes have generally been used for taxonomic purposes and for possible clues to the evolution of related groups of species. Unfortunately, this effort has not yet been as successful as similar studies in plant species. The somatic metaphase chromosomes of man, for example, are not easily distinguished from each other by their total length, arm ratio, or unusual topological markers such as heterochromatic segments, prominent chromomeres, or unique satellites (Fig. 8-1). According to the standard procedure, the metaphase chromosomes are serially numbered from the longest to the shortest, and morphologically similar chromosomes are placed within groups identified by letters. Chromosomes in different groups can be readily distinguished from each other. Consequently, the chromosomes within a group are often identified by the appropriate letter rather than its arbitrary number. The identification of each chromosome in man by some routine procedure is currently receiving considerable attention.

The meiotic chromosomes are obtained by sacrificing the animal or by biopsy, but the relatively high chromosome num-

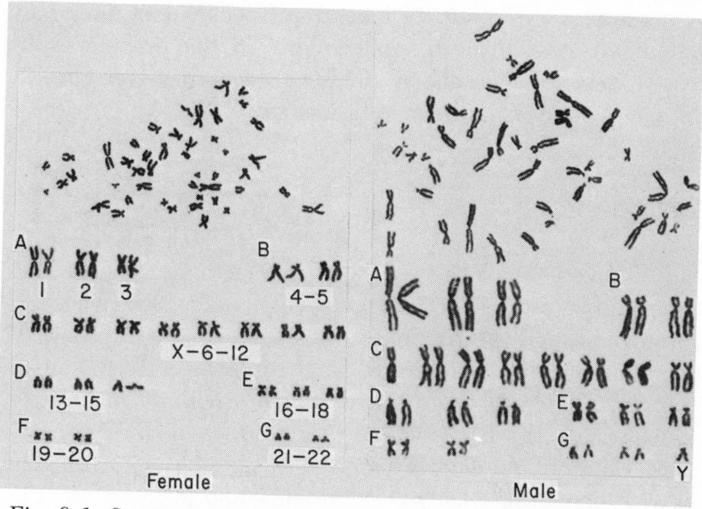

Fig. 8–1. Somatic metaphase chromosomes of a normal female and normal male arranged according to the London (1963) system to yield the karyotypes. (*Photographs courtesy of Dr. Janet Rowley.*)

ber usually frustrates attempts to detect aberrations at pachytene. Chromosomal associations can be observed at diakinesis or metaphase I, and a dicentric chromatid bridge and acentric fragment can be seen at anaphase I, indicating a heterozygous paracentric inversion. Somatic metaphase chromosomes, however, can reveal a gross deficiency when one chromosome is clearly shorter than its homolog and a fortunate pericentric inversion would yield a different arm ratio for one chromosome. A heterozygous reciprocal translocation could be detected by the altered morphology of two nonhomologous chromosomes. Although mammalian chromosomes are not exempt from structural aberrations, the small number of progeny reduces the opportunity to find aberrant chromosomes, and aberrant genetic ratios are not likely to provide clues. Mouse cytogeneticists can raise large populations of offspring from parents exposed to ionizing radiation to produce chromosomal aberrations.

Whenever aneuploidy is responsible for an altered phenotype, the loss or gain of a chromosome is easily determined by a chromosome count. In laboratory species with numerous mutant alleles, unexpected phenotypes in the progeny from crosses between parents of different genotypes can furnish clues to meiotic or mitotic nondisjunction.

CYTOGENETICS

The discovery of chromosome diseases in man stimulated a a surge of interest in human cytogenetics. In 1956, Tjio and Levan published the results of a cytological study of cells of lung tissue, grown in culture, from explants of four human embryos. The cultures furnished 265 cells with 46 metaphase chromosomes, the first accurate count for man. In 1959, Lejeune, Gautier, and Turpin reported 47 chromosomes for nine cases of mongolism (Down's syndrome), and in each case, the extra chromosome was identified as one of the smallest (group G) in the karyotype (Fig. 8-2). This discovery was soon confirmed and raised serious questions concerning the identification of extra chromosomes by different investigators. A standard karyotype was adopted by international conferences to avoid confusion.

An infant chimpanzee with the clinical and behavioral features of Down's syndrome was examined cytologically (Fig. 8-3) and found to be trisomic for a small acrocentric chromosome (McClure et al., 1969). Although the chimpanzee has 48 chromosomes, the association between the primary trisomic and the syndrome in man and chimpanzee suggests that the chromosome may contain similar genetic information in both primates. The discovery of one chromosomal disease in a primate suggests that a survey of poorly developing infant primates may reveal other chromosomal diseases.

Polani et al. (1960) reported a mongoloid child with 46 rather than the expected 47 chromosomes. A detailed study of the karyotype indicated four and not the anticipated five chromosomes in the G group, five and not six chromosomes in the D group, and the presence of an unusual chromosome

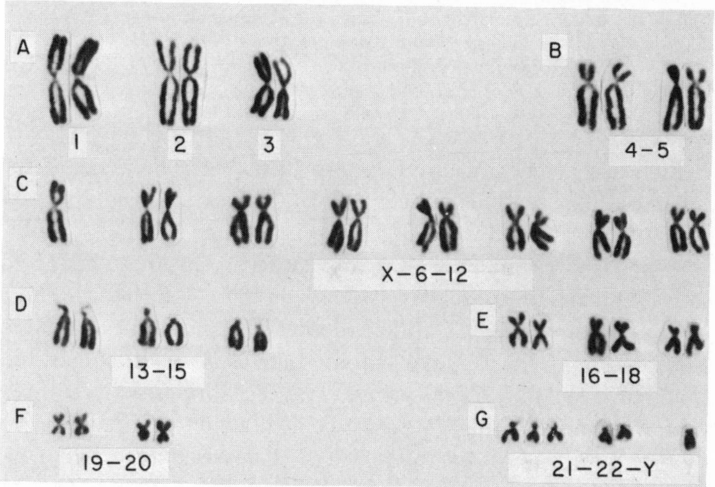

Fig. 8–2. Karyotype of a male exhibiting Down's syndrome and extra chromosome in group G, presumably chromosome 21. (*Photograph courtesy of Dr. Janet Rowley.*)

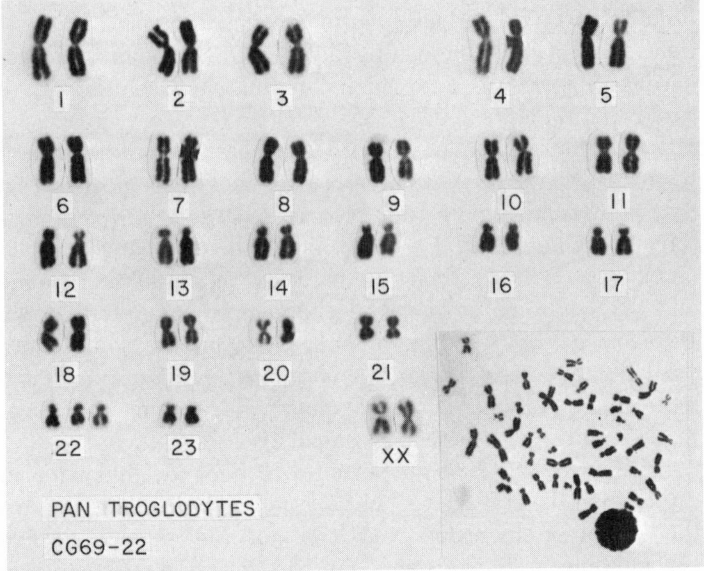

Fig. 8–3. Karyotype of a female chimpanzee with the clinical, behavioral, and cytogenetic characteristics of Down's syndrome. The extra small acrocentric chromosome has been arbitrarily designated as number 22. (*Photograph courtesy of Dr. H. M. McClure, Yerkes Regional Primate Research Center, Emory University; McClure et al., 1969.*)

which did not fit into the C group (Fig. 8-4). To account for all these observations, a reciprocal translocation was assumed to have involved chromosomes 15 (21^{15}) and 21 (15^{21}), yielding one short and one long translocation chromosome (Fig. 8-5). The loss of the small translocation indicated that the short arm of chromosomes 15 and 21 did not contain essential genetic information. The atypical chromosome was identified as the long translocation chromosome. The distribution of the three chromosomes during meiosis gave a mongoloid with 46 chromosomes ($2n - 15 + 15^{21}$). Consequently, the syndrome was caused by the duplication of the long arm of chromosome 21 and not by the duplication of the entire chromosome.

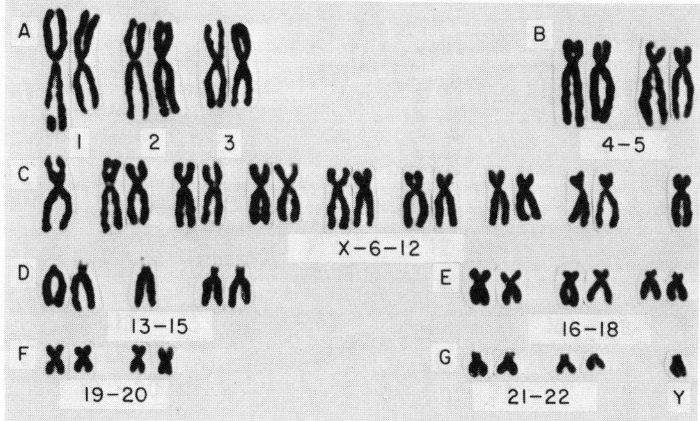

Fig. 8–4. Karyotype of a male ($2n = 46$) with Down's syndrome. Note the absence of one chromosome in group D and an extra chromosome (D/G translocation chromosome) lacking a homolog at the end of group C. (*Photograph courtesy of Dr. Janet Rowley.*)

A cytological examination of sisters who had translocation mongoloid offspring indicated that the mothers were apparently monosomic (Fig. 8-5). Their karyotypes and that of the grandmother indicated that these females lacked the short translocation chromosome and had the long translocation chromosome, thereby simulating a monosomic condition. These normal individuals are carriers who can produce mongoloid offspring with a relatively high frequency as well as carrier offspring.

While Down's syndrome is the most commonly encountered disease caused by a primary trisome, other autosomal primary trisomics have been found for groups B, D, and E. In each case, the primary trisomic exhibited a characteristic syndrome, usually involving mental retardation and atypical or abnormal development. The maximum number of 22 different autosomal primary trisomics may not be detectable because certain trisomes might have so drastic an impact on the development of the embryo or fetus that spontaneous abortions, or miscarriages, occur (Fig. 8-6).

Although a deficiency for the short arm of chromosome 15

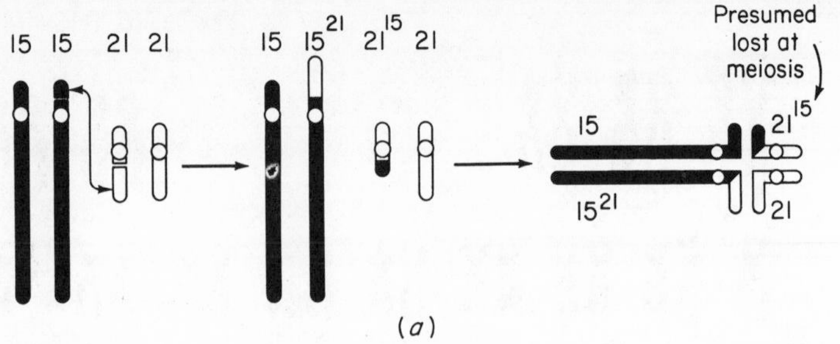

(a)

Original heterozygote			
15 21	15^{21}	15	15^{21} 21
15,15,21,21 Basic homozygote 2n=46	15^{21},15,21 Translocation carrier of normal phenotype 2n=45	15,15,21 Not recovered (presumed inviable) 2n=45	15,15^{21},21,21 Translocation Mongol 2n=46

(b)

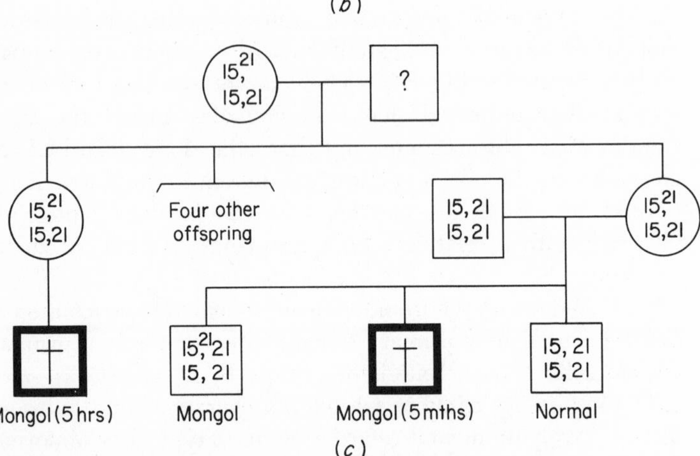

(c)

Fig. 8–5. The origin of translocation mongolism. (*a*) Reciprocal translocation involving chromosomes 15 and 21. (*b*) Gametes produced by the original translocation heterozygote. (*c*) Pedigree for a family yielding translocation mongolism. The four other offspring in the first generation died within 1 day to 15 months, and all displayed various bronchial disorders. (*Lewis and John, 1963; after Carter et al., 1960.*)

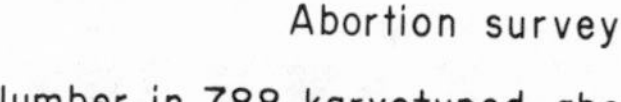

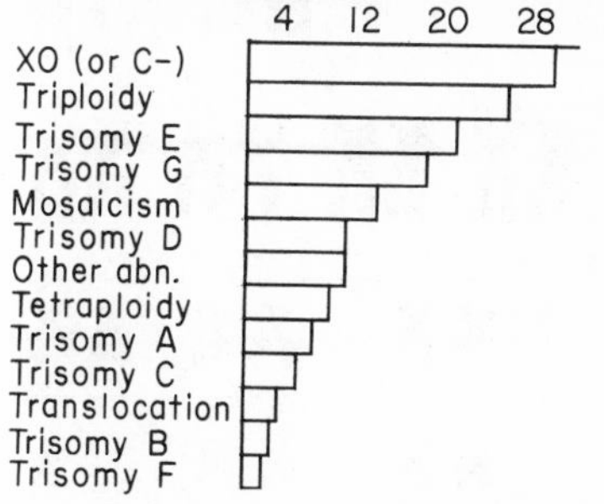

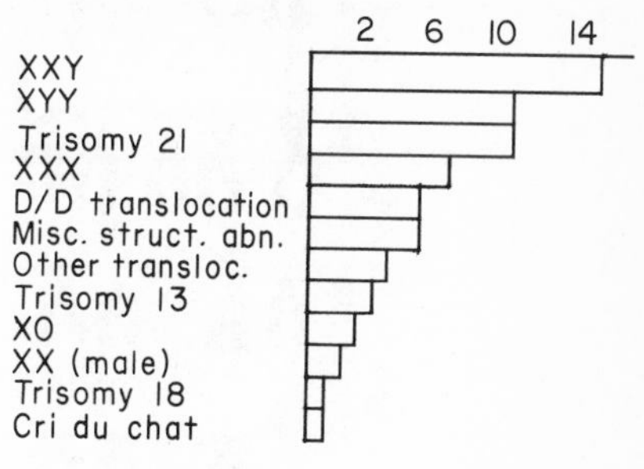

Fig. 8-6. Frequencies of different types of chromosomal aberrations detected in a survey of abortions and newborns. (*Courtesy of Dr. Janet Rowley.*)

or 21 does not produce an altered phenotype, a gross deficiency in the short arm of chromosome 5 is responsible for the *cri du chat* syndrome in infants (Fig. 8-7). The discovery of other deficiencies will depend on the constant association between an abnormal phenotype and a gross alteration in the length of a particular chromosome.

SEX CHROMOSOMES

Sex-linked genes were found in man before the sex chromosomes were identified cytologically. The X chromosome closely resembles the chromosomes in group C, but the Y chromosome is obviously different in morphology from any autosome. The role of these chromosomes in sex determination was unknown until the chromosomal basis of two syndromes was established. These syndromes had been described before the development of techniques to count and describe mammalian chromosomes. In Turner's syndrome, found only in females, the abnormal individuals exhibit faulty or incomplete ovarian development, sex hormonal imbalance, low-set ears, short stature, and a num-

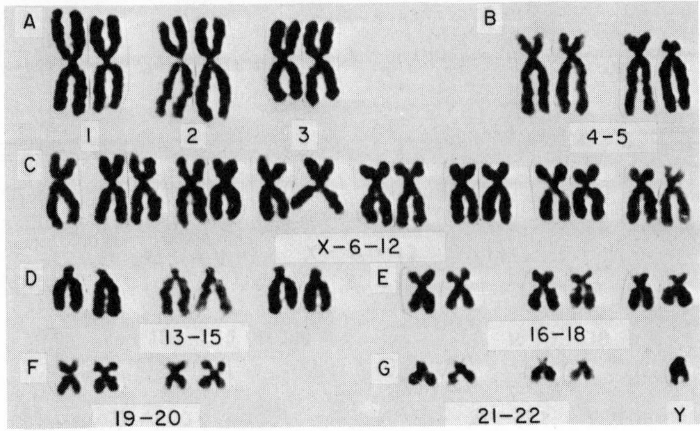

Fig. 8–7. Karyotype of an infant exhibiting the *cri du chat* syndrome. Note the difference in length of the short arm in one chromosome 5. (*Photograph courtesy of Dr. Janet Rowley.*)

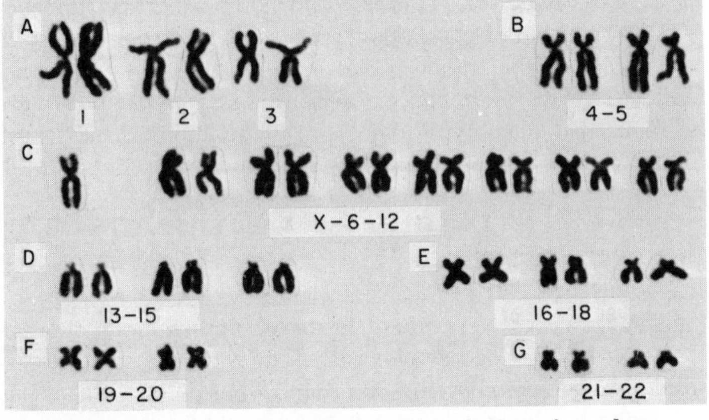

Fig. 8–8. Karotype of a female ($2n = 45$) with Turner's syndrome. (*Photograph courtesy of Dr. Janet Rowley.*)

ber of other apparently unrelated morphological alterations. In Klinefelter's syndrome, found only in males, the abnormal individuals are usually mentally retarded and have defectively developed testicles and a tendency to secondary female sexual characteristics.

Females with Turner's syndrome are monosomic, XO (Fig. 8-8), and males with Klinefelter's syndrome are usually trisomic, XXY (Fig. 8-9). Surveys of individuals with Klinefelter's syndrome revealed different numbers of X or Y chromosomes and always at least one Y chromosome (Table 8-1). The mental retardation associated with this syndrome becomes increasingly severe with added X chromosomes. Aneuploidy for the sex chromosomes has provided a number of interesting observations. The Y chromosome is male-determining, and its absence is female-determining in man and presumably in other mammalian species. In *D. melanogaster*, sex determination had been related to a ratio between the number of X chromosomes and sets of autosomes, and the Y chromosome was not involved. It should be noted that the Y chromosome in a dioecious plant species (*Melandrium album*) had already been established as the sex-determining chromosome.

The restriction of Turner's syndrome to females is directly related to the absence of the Y chromosome; Klinefelter's syndrome occurs in males because at least one Y chromosome is present. While the Y chromosome in man and in mouse appears to lack alleles for loci on the X chromosome, the Y

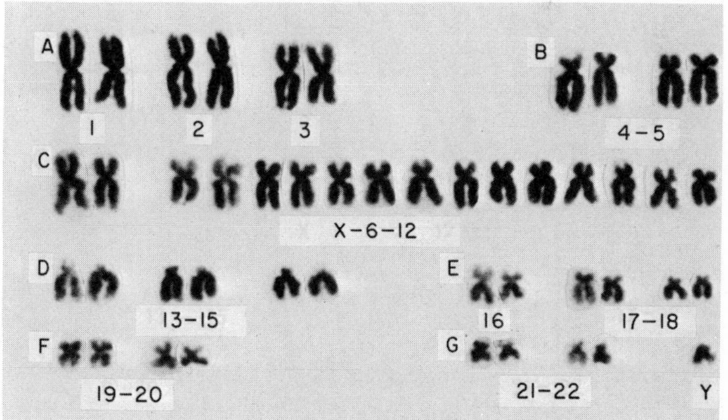

Fig. 8–9. Karyotype of a male (XXY) with Klinefelter's syndrome. The two X chromosomes appear in group C. (*Photograph courtesy of Dr. Janet Rowley.*)

TABLE 8-1

Aneuploidy for the Sex Chromosomes of *Homo sapiens*

Number of Y	Number of X Chromosomes				
Chromosomes	1	2	3	4	5
0	XO—45 Monosomic 0 Barr	XX—46 Disomic 1 Barr	XXX—47 Trisomic 2 Barrs	XXXX—48 Tetrasomic 3 Barrs	XXXXX—49 Pentasomic 4 Barrs
1	XY—46 Disomic 0 Barr	XXY—47 Trisomic 1 Barr	XXXY—48 Tetrasomic 2 Barrs	XXXXY—49 Pentasomic 3 Barrs	
2	XYY—47 Trisomic 0 Barr	XXYY—48 Tetrasomic 1 Barr	XXXYY—49 Pentasomic 2 Barrs		

chromosome of man cannot be viewed as genetically inert. The XO constitution in Turner's syndrome would be equivalent to normal males with XY if the heterochromatic Y chromosome were truly inert. While the Y chromosome is responsible for maleness, it has not yet been established that the genes responsible for the normal developmental steps leading to the primary sex differences are later repressed by the heterochromatization of this chromosome. Man differs from mouse in that the XO females in mice appear to be normal and fertile unlike XO human females.

Mitotic or meiotic nondisjunction of the sex chromosomes produces chromosome mosaics or gametes, respectively, with more or less than one X or Y chromosome. Thus, individuals with up to five sex chromosomes have been found, in contrast to the absence of autosomal tetrasomics or pentasomics. These observations suggest that the X or Y chromosomes might have other unusual characteristics to distinguish them from autosomes.

SEX CHROMATIN

Barr and Bertram (1949) reported an interesting example of sexual dimorphism in cats: a chromatinic body apparently adhering to the inner surface of the nuclear membrane of the resting nucleus in the neural cells of females but not of males (Fig. 8-10). In time, these observations were extended to the cells of other tissues for a number of mammalian species. The chromatinic body, termed *sex chromatin* or a *Barr body,* has played a significant role in understanding the cytogenetics of the mammalian X chromosome. The number of Barr bodies in the resting nucleus of mammalian cells is equal to 1 less than the number of X chromosomes.

The Barr body in man is approximately 1 micron in diameter. The nuclei in XX females have one Barr body and in XY males none. Cells from females (XO) with Turner's syndrome lack a Barr body, thereby simulating the cells of nor-

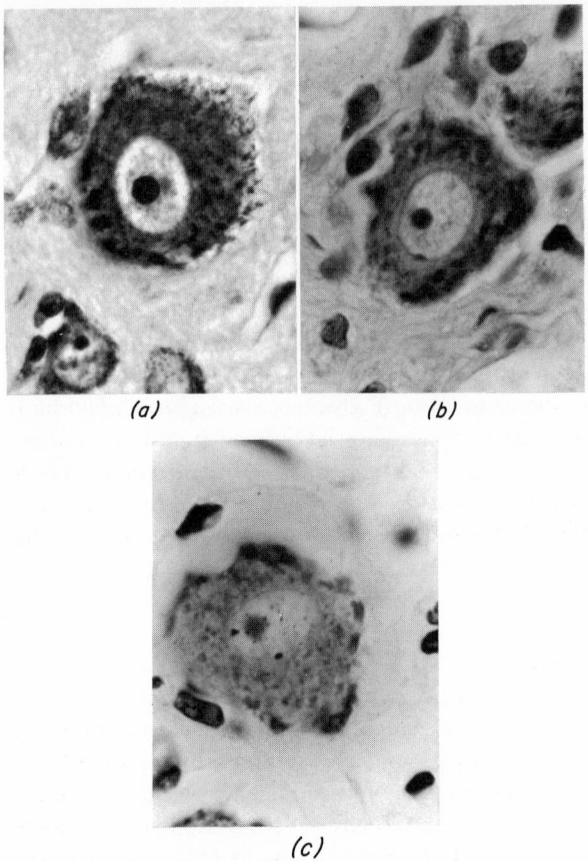

Fig. 8–10. Nuclei of neurons from (*a*) a normal male (XY) lacking a Barr body, (*b*) a normal female (XX) with one Barr body at 7 o'clock, and (*c*) a male (XXXY) with two Barr bodies, one at 5 o'clock and the other at 9 o'clock. (*Photographs courtesy of Dr. M. C. Barr.*)

mal males. Cells from XXX or XXXX females have two or three Barr bodies, respectively. Finally, males with Klinefelter's syndrome have cells with one or more Barr bodies, depending on the number of X chromosomes. The examination of cells from buccal scrapings for the number of Barr bodies has become a useful diagnostic tool in confirming Turner's syndrome, Klinefelter's syndrome, or multi-X syndromes and provides a rational basis for getting a chromosome count.

THE INACTIVE-X-CHROMOSOME HYPOTHESIS

Dominant sex-linked mutations for coat color in the mouse produce different phenotypes in the male and female. For example, heterozygous Mottled females (*Mo* +) have a variegated coat with patches of mutant or wild-type color while the hemizygous male (*Mo* Y) or monosomic female (*Mo* O) has a uniform coat. One explanation postulated an inactivation of the X chromosome, and the original hypothesis (Lyon, 1961; Russell, 1961) has been modified to accommodate a number of recent observations.

The inactive-X-chromosome hypothesis rests on a number of assumptions. During early embryogenesis and after the commitment to normal sexual development, either one of the X chromosomes of the female is inactivated, becomes heterochromatic, and appears in the resting nucleus as a Barr body. Clones from a cell with an inactivated X chromosome include this chromosome, which remains inactivated in successive mitotic divisions. In males or females with two or more X chromosomes, all but one X chromosome are inactivated and become heterochromatic. Finally, the inactivation of the X chromosome is restricted to somatic cells.

The variegation of heterozygous Mottled (*Mo mo*) female mice is readily explained by the inactive-X-chromosome hypothesis. In the cell with an inactive X chromosome carrying the dominant allele, the recessive phenotype appears because the dominant allele does not function. Additional experimental support for the hypothesis has come from studies with electrophoretic variants of glucose-6-phosphate dehydrogenase (G6PD) in man (Davidson et al., 1963). The alleles determining the variant enzymes are in the X chromosome. Although skin extracts from males yield only one site of G6PD activity, similar extracts from heterozygous females give two sites of activity. When individual skin cells from these females are grown in culture, the extracts from these tissue cultures yield one or the other site of activity but not both. The two sites of G6PD activity from the original skin extract were from a mixture of cells with different inactivated X chromosomes, and the extract from the tissue cultures contained the one enzyme determined by the allele in the active X chromosome.

Attempts to relate heterochromatin to the functional characteristics of genes have been successful to a limited degree (Baker, 1968). The cyclic events in the chromosomes during nuclear division seem to differ for segments which appear to be heterochromatic or euchromatic, suggesting that the difference might be related to the status rather than to the composition of the chromosomal material. For example, euchromatin generally replicates early in the S period of the mitotic cycle and heterochromatin at the end of this period. The inactivated X chromosome is heterochromatic and replicates later than the autosomes and the active X chromosomes.

While heterochromatic segments or chromosomes generally are considered to be devoid of Mendelian genes, heterochromatin can have an effect on gene action or chromosome behavior. An increase in the number of the heterochromatic supernumerary chromosomes in plant species results in reduced fertility or viability. In maize, a heterochromatic terminal segment in chromosome 10 is responsible for the neocentric activity of the heterochromatic knobs in the chromosomes. Mendelian genes are known to be nonfunctional in the inactivated X chromosome of mouse (Mottled) and man (glucose-6-phosphate dehydrogenase variant). Female mice heterozygous for an autosome–X chromosome reciprocal translocation and autosomal alleles for coat color are often variegated. In these cases, the inactivation extends from the X chromosomal segment to the autosomal segment with the dominant allele rendering this allele nonfunctional and permitting the recessive phenotype to be expressed. The spreading effect of the inactivation for the X chromosomal segment does not necessarily extend to the end of the autosomal segment in different translocation chromosomes (Russell, 1963).

The current interpretations of gene activity are couched in terms of the repression or derepression of genes in specific cells, tissues, or stages of development. If heterochromatin were the visible manifestation of the repression of certain loci, no further role would be played by these loci. On the other hand, heterochromatin may have some other function unrelated to the genes which had previously been active. The facultatively heterochromatic mammalian X chromosomes are not fully inert, particularly in man, with respect to the one X chromosome of normal females (XX) or to the several X chromosomes in the multi-X females or males. For example, a quantitative relation between the number of X chromosomes and the degree of brain malfunction has been observed for Klinefelter's syndrome.

The mammalian Y chromosome is considered to be mostly heterochromatic, and there is no evidence of alleles for loci in the X chromosome. While the Y chromosome is responsible for maleness, data are not available to indicate that the genes responsible for the normal developmental steps leading to the primary sex differences are later derepressed by the heterochromatization of this chromosome. The absence of the Y or X chromosome in the human female results in Turner's syndrome, but the XX female with a heterochromatic X chromosome and the XY male with the heterochromatic Y chromosome are normale. On the other hand, the female XO mouse is apparently normal and fertile.

The heterochromatization and presumably inactivation of chromosomes have been reported for the mealybugs and scale insects (Brown, 1966). In the males, a haploid set of chromosomes becomes heterochromatic in the nuclei of the cells of the developing embryo at the blastula stage. The heterochromatic haploid set has been shown to be the paternal contribution to the zygotic nucleus.

GONOSOMIC MOSAICISM

Ohno (1964) reported an interesting situation in the creeping vole (*Microtus oregoni*) related to the sex chromosomes. Although the male zygote is XY and the female zygote XO, the somatic and germinal cells differ in chromosome number and type of sex chromosomes. Mitotic nondisjunction in the primordial germinal cells of the developing male embryo is responsible for the loss of the X chromosome. During spermatogenesis, the Y chromosome goes to one pole during anaphase I and divides mitotically during the second meiotic division, so that only two of the four sperm will have one Y chromosome and the other two have no sex chromosomes. In the XO female embryo, a mitotic nondisjunction of the X chromosome is responsible for germinal cells with XX chromosomes. The female has oocytes with XX chromosomes which behave normally during oogenesis, and all the eggs have one X chromosome. The difference in number and type of sex chromosomes of the somatic and germinal cells of the male and female creeping vole has been termed *gonosomic mosaicism.*

With only a single X chromosome in the somatic cells of the female creeping vole, there is no need to inactivate one X chromosome as in other mammalian females with XX chromosomes in the somatic cells. Creeping voles of either sex with more than the normal number of X chromosomes should be interesting experimental material to determine whether this species has retained the mechanism for inactivating the X chromosomes.

REFERENCES

BAKER, W. K. 1968. Position-effect variegation. Advances Genet. 14:133–169.

BARR, M. L., AND L. F. BERTRAM. 1949. A morphological distinction between neurones of the male and female and the behavior of the nucleolar satellite during accelerated nucleoprotein synthesis. Nature 163:676–677.

BROWN, S. W. 1966. Heterochromatin. Science 151:417–425.

CARTER, C. O., J. L. HAMERTON, P. POLANI, A. GUNLAP, AND S. D. V. WELLER. 1960. Chromosome translocation as a cause of familial mongolism. Lancet 2:678–680.

DAVIDSON, R. G., H. M. NITOWSKY, AND B. CHILDS. 1963. Demonstration of two populations of cells in the human female heterozygous for glucose-6-phosphate dehydrogenase variants. Proc. Nat. Acad. Sci. 50:481–485.

LEJEUNE, J., M. GAUTIER, AND R. TURPIN. 1959. Étude des chromosomes somatiques de neuf enfants mogoliens. Compt. Rend. Acad. Sci. 248:1721–1722.

LEWIS, K. R., AND B. JOHN. 1963. Chromosome marker. J. & A. Churchill, London.

LYON, M. 1961. Gene action in the X-chromosome of the mouse (*Mus musculus* L.). Nature 190:372–373.

MCCLURE, H. M., K. H. BELDEN, W. A. PIEPER, AND C. B. JACOBSON. 1969. Autosomal trisomy in a chimpanzee: resemblance to Down's syndrome. Science 165:1010–1011.

OHNO, S. 1964. Restoration of XX-öogonia in XO females of *Microtus oregoni*, pp. 40–42. *In* Congenital malformations. The International Medical Congress, New York.

POLANI, P. E., J. H. BRIGGS, C. E. FORD, C. M. CLARK, AND J. M. BERG. 1960. A mongol girl with 46 chromosomes. Lancet 1:721–724.

RUSSELL, L. B. 1961. Genetics of mammalian sex chromosomes. Science 133:1795–1803.

———. 1963. Mammalian X-chromosome action: inactivation limited in spread and in region of origin. Science 140:976–978.

TJIO, J. H., AND A. LEVAN. 1956. Chromosome number of man. Hereditas 42:1–6.

SUPPLEMENTARY REFERENCES

BENIRSCHKE, K. (ed.). 1969. Comparative mammalian cytogenetics. Springer-Verlag, New York.

Carr, D. H. 1969. Chromosomal abnormalities in clinical medicine. Progr. Med. Genet. 6:1–61.

Mittwoch, U. 1967. Sex chromosomes. Academic, New York.

Turpin, R., and J. Lejeune. 1969. Human afflictions and chromosomal aberrations. Pergamon, New York.

9 CYTOGENETICS AND SPECIATION

The chromosome number is as important a datum for a species as any other characteristic deemed significantly stable to merit taxonomic significance. The cytotaxonomic surveys of plant species during the first two decades of this century indicated that populations of one species generally have the same chromosome number but genera can include species with different numbers. Furthermore, such genera are significantly more frequent in higher plants than in animals, and in many plant genera, the different chromosome numbers for the species are multiples of a basic number. Winge (1917) first proposed an explanation for these observations in plant genera: diploid species yield interspecific hybrids with reduced or complete sterility which produce seeds as a result of a spontaneous chromosome doubling either in somatic or meiotic cells of the hybrids and thereby produce tetraploid populations which can evolve into tetraploid species. This process may be repeated in hybrids between diploid and tetraploid species to produce hexaploid populations which eventually yield hexaploid species. In 1913, Federley already had reported that a fertile hybrid between diploid species, *Pygaera curtula* and *P. anachoreta,* in a butterfly genus gave diploid gametes and noted that new species might arise from such hybrids.

Pellew and Dunham (1916) provided cytological evidence to support Winge's hypothesis from a study of *Primula kewensis.* A spontaneous interspecific hybrid between *P. floribunda* ($n = 9$) and *P. verticillata* ($n = 9$) appeared in a population of the former species at Kew, England, in 1900. The sterile hybrid was propagated vegetatively and seed discovered in 1905 gave a fertile plant whose fertile uniform progeny differed not only in their chromosome number but also in their morphology when compared with the parental species. The unique status of these progeny with 18 bivalents was recognized by proposing the specific epithet *P. kewensis.* Clausen and Goodspeed (1925) crossed *Nicotiana tabacum* ($n = 24$) and *N. glutinosa* ($n = 12$) and obtained a population of sterile interspecific hybrids, except for one robust fertile plant. The progeny from this hybrid were fertile, relatively uniform, and had 36 bivalents; they were viewed as experimental evidence supporting Winge's hypothesis on the origin of polyploid species.

CLASSIFICATION OF POLYPLOID SPECIES

Cytological studies of polyploid species and of interspecific hybrids had indicated that the classification of polyploid species would not be a simple matter. A confusing array of terms was generated by cytotaxonomists in their efforts to account for the chromosome associations observed both in the interspecific hybrids and in the polyploid species. The genome or set of chromosomes (represented by X) in the gametic nucleus of the basic diploid species of a genus was used as the unit in discussing the origin of polyploid species. For example, the genus *Triticum* includes species with a gametic chromosome number of 7, 14, or 21, and in this genus, one genome has seven chromosomes ($X = 7$). Furthermore, the diploid (2X), tetraploid (4X), and hexaploid (6X) species have bivalents at diakinesis or metaphase I, suggesting that the tetraploid species have two different genomes and the hexaploid species have three different genomes in the gametic nucleus. It should be emphasized that X and n are not interchangeable symbols.

Polyploid species are classified as autopolyploids, segmental allopolyploids, true or genomic allopolyploids, or autoallopolyploids. The classification considers the origin and the relation of the genomes (Fig. 9-1). According to Stebbins (1950),

> The first two occur in nature predominantly or entirely at the level of triploidy or tetraploidy; true allopolyploidy can occur at any level from tetraploidy upwards while autoallopolyploidy is confined to hexaploidy and higher levels of polyploidy. The term amphiploid, coined by Clausen, Keck and Hiesey (1945), is suggested as a collective term to cover all types of polyploidy which have arisen after hybridization between two or more diploid species separated by barriers of hybrid sterility.

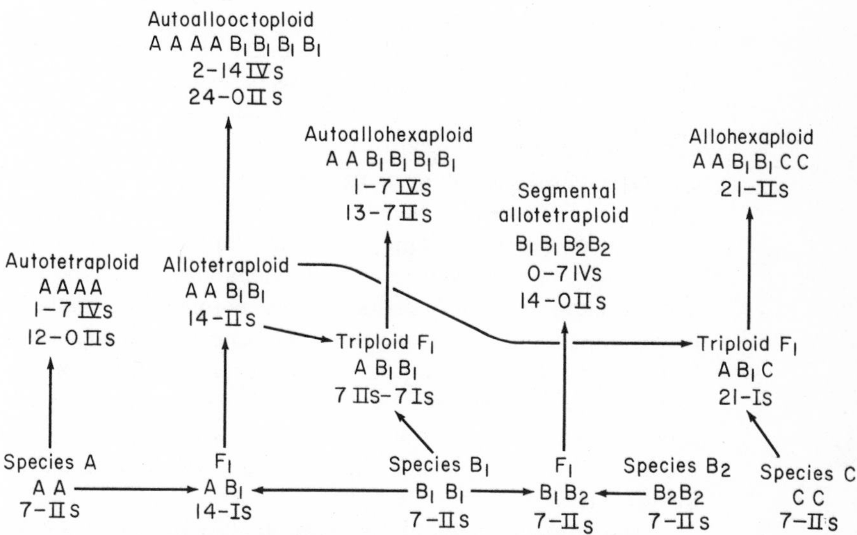

Fig. 9–1. Idealized diagram illustrating the interrelationships, genome constitution, and origin of typical autopolyploids, allopolyploids, segmental allopolyploids, and autoallopolyploids. (*Stebbins, 1950.*)

> It therefore includes segmental allopolyploids, true allopolyploids, and autoallopolyploids plus aneuploids which have arisen from hybridization between two species belonging to an aneuploid series with lower numbers. . . .

Autotetraploid species with four homologous genomes are derived from a single member of a diploid species and exhibit quadrivalents at diakinesis or metaphase I. Tetraploid species with quadrivalents, however, can also be produced by diploid hybrids between closely related species. The deciding factor is the fertility or sterility of the diploid progenitor, information that is usually unavailable. Consequently, the acceptable candidates for autotetraploid species have been restricted to monotypic genera.

Segmental allotetraploids are derived from poorly fertile or sterile diploid interspecific hybrids with bivalents or bivalents and univalents. Although the genomes of the parental species are not homologous, chromosomes in each genome share homologous segments to account for the bivalents in the hybrids and for the quadrivalents in the tetraploid species. When the diploid interspecific hybrids have only bivalents, the genomes from the parental species might seem to be homologous, but the distribution of the chromosomes during sporogenesis yields aborted or malfunctioning spores or gametophytes.

True or genomic allopolyploids are derived from interspecific hybrids between species with nonhomologous genomes. An allotetraploid eventually results from a chromosome doubling in the diploid hybrid with univalents and the allohexaploid from a chromosome doubling in the triploid hybrid with univalents from a cross between a diploid and an allotetraploid species. The fertility of the allopolyploid is directly related to the bivalents, in which each chromosome of the hybrids has an identical homolog.

The autoallopolyploids are hexaploid or octoploid species with bivalents and quadrivalents at diakinesis or metaphase I. The failure to form a quadrivalent may result in a trivalent and univalent or a bivalent and two univalents. Chromosome doubling in a triploid interspecific hybrid between an allotetraploid and a diploid species with one genome already present in the allotetraploid yields an autoallohexaploid; chromosome doubling in an allotetraploid produces an autoallooctoploid.

GENOME ANALYSIS

The usual procedure in genome analysis requires a cytological study of the species in the genus and a survey of the crossability of these species to obtain interspecific hybrids. The cytological survey establishes the presence or absence of a polyploid species and, when a polyploid series is present, the basic chromosome number (X) of the genomes. The chromosome associations and the fertility or sterility of the diploid interspecific hybrids furnish information on the homology, degree of homology, or nonhomology of the genomes in the diploid species. The chromosome associations in the hybrids between the tetraploid and diploid species also provide information on the homology or nonhomology of the genomes in the tetraploid and diploid species. The interpretation of chromosome associations in the interspecific hybrids rests on certain assumptions pertinent to the number of chiasmata at diakinesis or metaphase I. Finally, the assumption that all genomes in the polyploid species are present in the available diploid species must include the possibility that a diploid species might have contributed its genome to a polyploid species and then become extinct or lost to taxonomists.

The correct interpretation of the chromosome associations in a polyploid species is dependent on the chromosome associations in the interspecific hybrid which was the progenitor of the polyploid species. The chromosome associations at diakinesis or metaphase I in the polyploid species or interspecific hybrids are determined by at least four factors: (1) the relative length of homologous segments in the chromosomes of different genomes, (2) the rate of terminalization of chiasmata from diplotene to diakinesis or metaphase I, (3) preferential pairing, and (4) genetic factors. For example, *Helianthus tuberosus* ($2n = 102$) and *H. annuus* ($2n = 34$) belong to a genus with a basic genome of 17 chromosomes; the former species is hexaploid and the latter diploid. The tetraploid interspecific hybrid has 34 bivalents at metaphase I (Kostoff, 1939). To account for these observations, *H. tuberosus* was assigned a genomic formula of $A_tA_tA_tA_tB_tB_t$, *H. annuus* B_aB_a, and the interspecific hybrid $A_tA_tB_tB_a$. The two A_t genomes in the interspecific hybrid paired and were either homologous or had chromosomes with homologous segments; the B_t and B_a genomes differed completely from the A genomes and were either homologous or had chromosomes with homologous segments. Consequently, *H. tuberosus* is either an autoallopolyploid or a segmental allopolyploid.

Chromosome pairing in polyploid species and interspecific hybrids has been viewed in terms of the origin of the bivalents in the hybrids. In *autosyndesis*, the chromosomes of the genomes contributed by one parental species form bivalents, as in the case of the A_tA_t genomes from *H. tuberosus*, and in *allosyndesis*, the chromosomes of genomes from different parental species form bivalents, as in the case of the B_t and B_a genomes from *H. tuberosus* and *H. annuus*, respectively. In segmental allotetraploids, the genomes are not completely homologous, and this situation is recognized by the genomic formula $B_1B_1B_2B_2$. The quadrivalent in the hybrid indicates an association of $B_1B_1B_2B_2$ chromosomes by the formation of chiasmata in homologous segments. The pairing of the B_1B_1 or of B_2B_2 chromosomes is termed *homogenetic association* and of B_1B_2 chromosomes, *heterogenetic association*. When structurally identical chromosomes pair preferentially, the potential quadrivalent is replaced by two bivalents (homogenetic association).

The classification of polyploid species by chromosome associations in the species and in the interspecific hybrids seems to be straightforward in theory but can become difficult or even arbitrary in practice. In many cases, the assignment of polyploid species to one of the four categories should be viewed with caution. The potential role of genetic factors and of preferential pairing in assessing the significance of chromosome pairing in the polyploid species cannot be overlooked.

The genus *Triticum* (X = 7), with diploid, tetraploid, and hexaploid species, is closely related to other genera in the family Triticineae, and intergeneric hybrids are obtained. The agronomically important hexaploid species *T. aestivum* has 21 bivalents, thereby appearing to be a genomic allohexaploid with three nonhomologous genomes (ABD). Interspecific and intergeneric hybrids provided the cytological evidence indicating the probable origin of this species. The A genome was contributed by a diploid species of *Triticum,* the B genome by *T.* (*Aegilops*) *squarrosa.* This genomic formula was acceptable until a cytological study of the 21 nullisomics in *T. aestivum* suggested that the presence of bivalents was not a valid basis for viewing this species as a genomic allohexaploid.

The morphological comparisons of nullisomic-tetrasomic combinations in *T. aestivum* indicated seven sets of three homoeologous chromosomes, one from each of the three genomes. Of the 21 nullisomics, only plants nullisomic for chromosome V had quadrivalents. The distribution and frequency of chiasmata in standard and nullisomic V plants were not significantly different, suggesting that chiasma formation was probably not responsible for the absence of quadrivalents in the standard plants. In reconstructing the probable origin of *T. aestivum,* an interspecific hybridization at the diploid level was proposed, and a spontaneous chromosome doubling in diploid hybrids produced a fertile tetraploid. The distribution of the chromosomes in the quadrivalents in the tetraploid during meiosis presumably was responsible for reduced fertility. A mutation suppressing the formation of quadrivalents and enhancing the frequency of bivalents increased fertility and therefore had selective value. Finally, a second hybridization between the tetraploid and a diploid species gave a sterile triploid, which, after chromosome doubling, produced a fertile hexaploid. In the final analysis, bivalents in polyploid species constitute presumptive evidence that these species are genomic allopolyploids, particularly when cytological data from interspecific hybrids are lacking.

The chromosomes from different genomes of closely related diploid species can form bivalents in the interspecific hybrids. In the genus *Collinsia* ($X = 7$), diploid species were involved in numerous interspecific hybrids which formed bivalents or interchange complexes at diakinesis or metaphase I. The cytological evidence suggested that the genomes of different species were homologous but the chromosomes had been repatterned by reciprocal translocation and paracentric inversions which became homozygous in each species. Autotetraploids had up to the maximum number of quadrivalents at diakinesis or metaphase I. In several amphiploids, the maximum number of quadrivalents was one or two, depending on the number of bivalents with structurally identical homologs in the corresponding diploid interspecific hybrids (Table 9-1). Each structurally altered chromosome in the diploid interspecific hybrids had identical homologs in the corresponding amphiploids, and these chromosomes preferentially paired to yield bivalents. For example, an interchange complex of four chromosomes and five bivalents in a diploid interspecific hybrid gave four bivalents and a maximum of five quadrivalents in the amphiploid. Furthermore, a bivalent with a standard and inversion homolog produced two bivalents in the amphiploid which did not display a dicentric chromatid bridge and acentric fragment at anaphase I. According to this interpretation of chromosome associations in the different amphiploids of *Collinsia,* a diploid interspecific hybrid in which all the bivalents had homologs with structural aberrations should yield an amphiploid with 14 bivalents. Hayhome and Garber (1968) obtained an amphiploid with 14 bivalents from a hybridization between *C. concolor* and a diploid F_2 plant from the interspecific hybrid *C. stricta* $\times$ *C. concolor.* In the genus *Collinsia,* amphiploids can be synthesized to simulate segmental or genomic allotetraploids even though the genomes are essentially homologous.

TABLE 9-1

Chromosome Associations in Four Diploid Interspecific Hybrids and the Corresponding Amphiploids in the Genus *Collinsia*

Diploid Interspecific Hybrids	Chromosome Associations, Metaphase I	Dicentric Bridges and Fragments, Anaphase I	Maximum Number of Quadrivalents in Amphiploids, Metaphase I
C. concolor × *C. corymbosa* *	$1\odot 6 + 4^{II}$	0	4
C. concolor × *C. sparsiflora*	$2\odot 4 + 3^{II}$	2	2
C. sparsiflora × *C. bartsiaefolia*	$1\odot 8 + 3^{II}$	2	1
C. solitaria × *C. heterophylla*	$1\odot 8 + 1\odot 4 + 1^{II}$	2	1

* Chromosome formulas for each species indicated an interchange complex of six chromosomes for the interspecific hybrid which had univalents and bivalents. The maximum number of four quadrivalents in the corresponding amphiploid supported the chromosome formula for these species.

Source: Garber and Dhillon, 1962.

DYSPLOIDY

A cytological survey of the species in a genus may reveal different chromosome numbers which are not multiples of a basic number. Genera with a dysploid series ($n = 3, 4, 5,$ or 6) are more common in animals than in plants. One explanation for dysploidy involves reciprocal translocations between nonhomologous chromosomes (Fig. 9-2). In these reciprocal translocations, one break occurs in the short arm and the other in the long arm of nonhomologous chromosomes, or both breaks are in nonhomologous centromeres. To reduce the chromosome number, the small translocation chromosome is lost during meiosis, and the long translocation chromosome with all the essential genetic information for both chromosomes is included in a gametic nucleus. When a fertilization involves these gametic nuclei, the zygote has a reduced chromosome number with all the essential genetic information. To increase the chromosome number, the first reciprocal translocation is followed by a second reciprocal translocation involving the small translocation chromosome and the long arm of a standard chromosome. In this case, the gametic nucleus has two chromosomes with all the genetic information formerly contained in one standard chromosome and one of the original standard chromosomes. The increase or decrease in chromosome number commences in a basic diploid species with a sufficient number of chromosomes to provide the necessary centromeres to account for the other species with the highest and lowest chromosome number in the dysploid series.

The prevalence of dysploid series in animal genera warrants comment. Natural interspecific hybrids are relatively uncommon in animals. Ecologic, geographic, and behavioral differences among related animal species militate against such hybrids. When chromosomal aberrations occur and eventually become homozygous in an animal species, hybrids between the standard and chromosomally different populations are either poorly fertile or sterile, depending on the type and number of such aberrations. If a reduction or an increase in chromosome

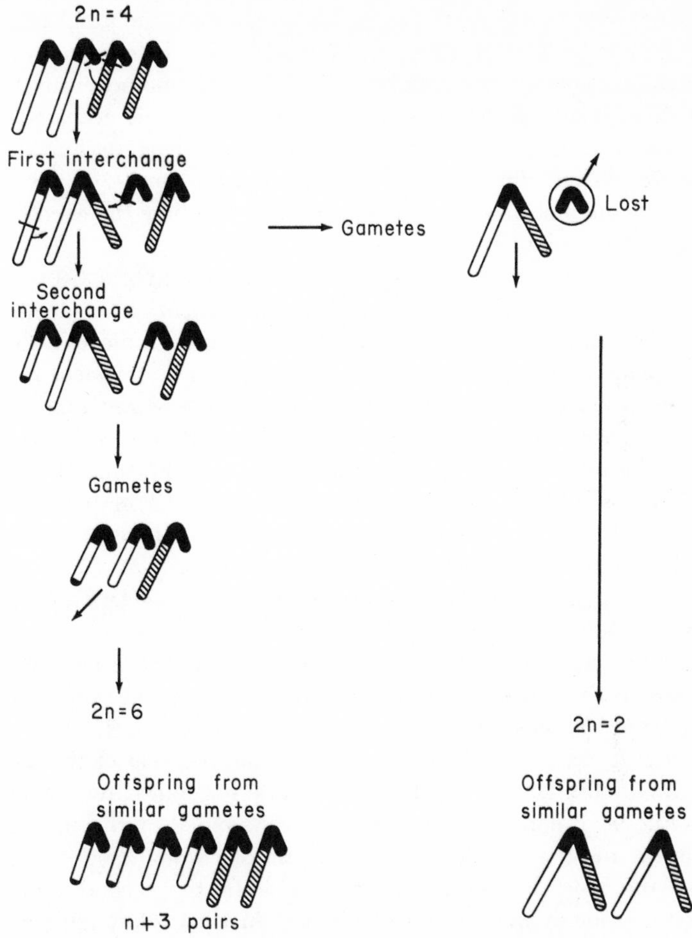

Fig. 9–2. The role of reciprocal translocations in increasing or decreasing chromosome numbers. The black segments are assumed to be genetically inert. (*Stebbins, 1950; after Darlington, 1937.*)

number were to occur, hybrids between populations with different chromosome numbers would be sterile. The absence of gene flow between these populations sets the stage for incipient speciation. Mutations altering mating behavior or adaptability to different environments reinforce the isolation of populations. The absence of a mating barrier between closely related species with different chromosome numbers can be demonstrated in the laboratory. In plants, there is no need to consider mating barriers to interspecific hybridization. Furthermore, only a single fertile amphiploid from sterile or poorly fertile interspecific hybrids is necessary to initiate the eventual emergence of a tetraploid species.

A cytogenetic analysis of an interspecific hybrid between *Crepis fuliginosa* ($n = 3$) and *Crepis neglecta* ($n = 4$) was undertaken because the taxonomic evidence indicated that *C. fuliginosa* might have been derived from *C. neglecta* or a closely related species (Tobgy, 1943). This investigation furnished the experimental evidence to support the explanation for the origin of a dysploid series in a genus. Fortunately, the chromosomes of each species can be identified in the interspecific hybrid, and the chromosome associations at metaphase I are readily interpreted (Fig. 9-3).

By noting the chromosome associations at metaphase I in the hybrids, the chromosomes were assigned to one of two groups: A_F A_N D_F D_N and B_N B_F C_N. In the first group, chromosomes D_F and D_N could pair in their long and short arms as could chromosomes A_F and A_N; the short arm of chromosome A could pair distally with the distal segment of the long arm of chromosome D_F. In an association of four chromosomes (Fig. 9-3), chromosomes A and D were joined by the chiasma in the distal segment of the short arm of chromosome A_N and of the long arm of chromosome D_F. In the second group, chromosomes B_F and B_N could pair in their short and long arms while one arm of chromosome C_N paired with the distal segment of the long arm of chromosome B_F. To account for these observations, a reciprocal translocation in *C. neglecta* was assumed to involve a distal segment of the long arm of chromosome B and a segment in one arm of chromosome C. The small translocation chromosome with the C centromere was lost. The long translocation chromosome with the B cen-

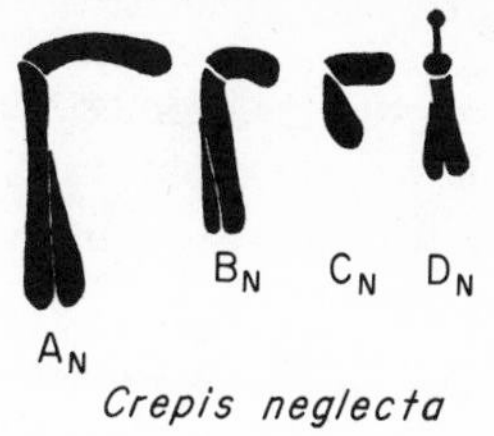

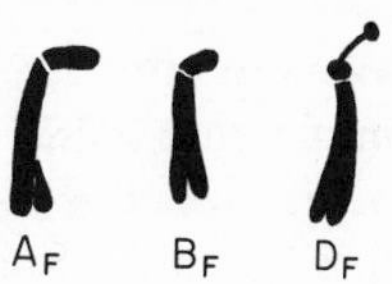

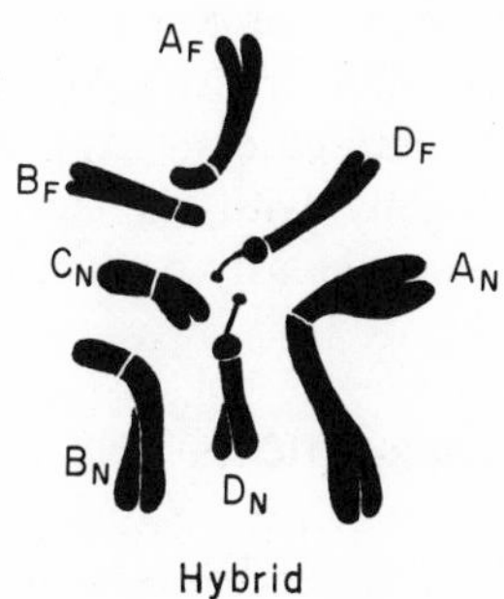

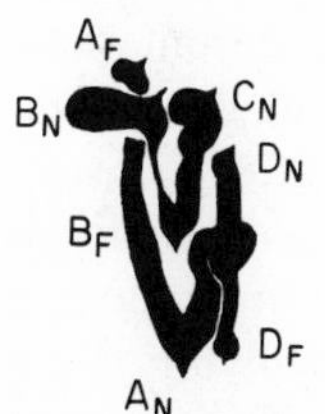

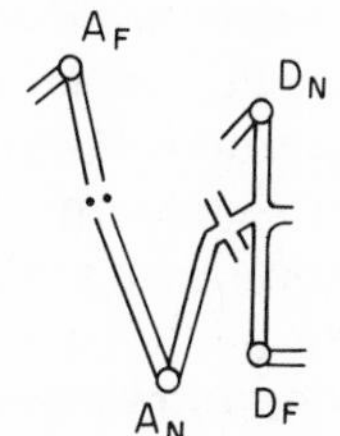

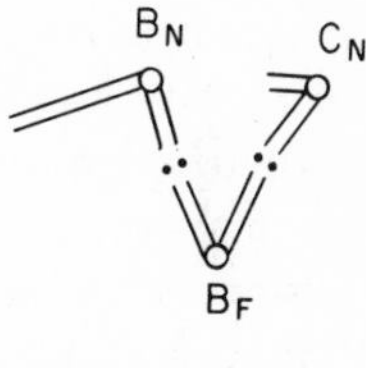

Fig. 9–3. *Upper diagrams:* somatic chromosomes of *Crepis neglecta* ($n = 4$) and *C. fuliginosa* ($n = 3$). *Middle diagram:* somatic chromosomes of the interspecific hybrid. *Lower diagrams:* significant chromosome associations at metaphase I in the interspecific hybrid and their interpretation. (*Tobgy, 1943.*)

tromere in *C. neglecta* was the ancestor of the B chromosome in *C. fuliginosa.* Moreover, a reciprocal translocation in *C. neglecta* between chromosomes A and D furnished the ancestors for chromosomes A and D in *C. fuliginosa.*

Dubinin (1936) used reciprocal translocations in *Drosophila melanogaster* ($2n = 8$) to obtain individuals with $2n = 6$. A reciprocal translocation between chromosome IV and the Y chromosome transferred all the essential genetic information from the small chromosome to the Y chromosome. Then an X-Y translocation transferred the chromosome IV information to an X chromosome. Appropriate matings from these flies gave progeny with six chromosomes and all the essential information previously present in eight chromosomes.

CHROMOSOMAL ABERRATIONS AND SPECIATION

Reciprocal translocations and inversions occur spontaneously and in time can become homozygous in viable, fertile individuals with altered chromosome morphology. For example, a pericentric inversion might change a metacentric chromosome to an extremely acrocentric one with no significant genetic information in the very short arm. In homozygotes for such chromosomes, a reciprocal translocation can lead to individuals with more or less than the original chromosome number and initiate a dysploid series. Intraspecific hybrids involving members of the original population and chromosomally aberrant population will usually be sterile. Once the populations are genetically isolated, the population with the new chromosomal organization has an opportunity to accumulate mutations which can lead to morphological and physiological changes associated with speciation.

The pseudolinkage from heterozygous reciprocal translocations is able to create favorable gene combinations which are maintained whenever the population has or develops the means to maintain these combinations. The directed orientation of interchange complexes and the pollen, egg, and zygotic lethals in *Oenothera* are examples of devices for exploiting the advantages of pseudolinkage.

Barriers to the flow of genes between populations of a species also arise by mutations which directly or indirectly

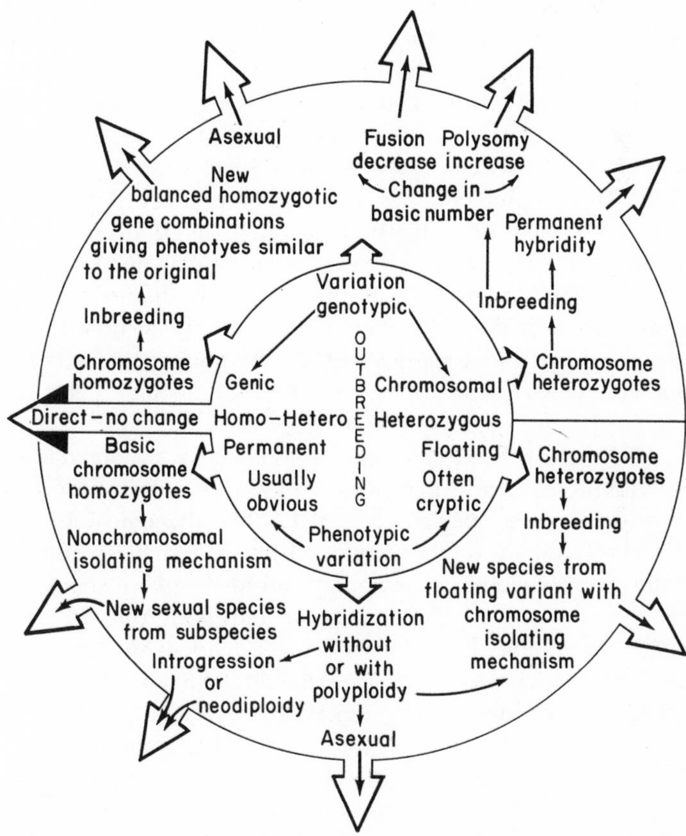

Fig. 9–4. The possible consequences of chromosomal changes in space and time as they relate to inbreeding, outbreeding, or asexual species. (*John and Lewis, 1963.*)

interfere with normal crossing or mating patterns. When mutations confer adaptability to environments different from those already occupied by most of the members of the species, the mutants are able to move from the original range of distribution and eventually become physically isolated. In plants, two species with overlapping distributions and opportunities for interspecific hybridization can produce reasonably fertile interspecific hybrids at the diploid level. The shuffling of the chromosomes or of chromosomal segments after crossing-over may eventually produce a third species. When the interspecific hybrid is poorly fertile to completely sterile, chromosome doubling restores fertility and usually furnishes a barrier to the flow of genes between the progenitor species and the amphiploid. The amphiploid may be able to produce populations with a characteristic distribution range, and, in time, the amphiploid emerges as a "new" species. Polyploid species, however, are less likely to produce radically different phenotypes than diploid species. The duplication of genetic information in the polyploid species restricts the segregation of spontaneous recessive mutations, and recessive homozygotes are slow to appear in the population.

The genetic and chromosomal changes in populations provide the cast for a play which may or may not succeed in producing a new species (Fig. 9-4). The dynamic interplay between the opportunities provided by such changes and the mode of reproduction are complex and involve both space and time. The absence of mating barriers and the fertility of spontaneous amphiploids in plants have provided an avenue of speciation which was effectively denied to animals. The genetic isolation of populations by mutations and intrachromosomal structural aberrations was exploited by both plants and animals. An understanding of the chromosomal devices exploited by populations to achieve genetic isolation has come from the cytogenetic analysis of intraspecific intrachromosomal, interchromosomal, and numerical aberrations.

REFERENCES

CLAUSEN, J., D. D. KECK, AND W. M. HIESEY. 1945. Experimental studies on the nature of species. II. Plant evolution through amphiploidy and autoploidy, with examples from the Madiinae. Carnegie Inst. Washington Pub. 564.

CLAUSEN, R. E., AND T. H. GOODSPEED. 1925. Interspecific hybridization in *Nicotiana.* II. A tetraploid *glutinosa-tabacum* hybrid: an experimental verification of Winge's hypothesis. Genetics 10:278–284.

DARLINGTON, C. D. 1937. Recent advances in cytology, 2d ed. Blakiston, Philadelphia.

DUBININ, N. P. 1936. Experimental alteration of the number of chromosome pairs in *Drosophila melanogaster.* Biol. Zentralbl. 3:719–736.

FEDERLEY, H. 1913. Das Verhalten der Chromosomen bei des Spermatogenese der Schmetterlinge *Pygaera anachoreta, curtula,* und *pigra* sowie einiger ihrer Bastarde. Z. Induktive Abstammungs- Vererbungslehre 9:1–110.

GARBER, E. D., AND T. S. DHILLON. 1962. The genus *Collinsia.* XVII. Preferential pairing in four amphidiploids and three triploid interspecific hybrids. Can. J. Genet. Cytol. 4:6–13.

HAYHOME, B. A., AND E. D. GARBER. 1968. The genus *Collinsia.* XXIX. Preferential pairing in diploid, triploid and tetraploid interspecific hybrids involving *C. stricta* × *C. concolor* and related species. Cytologia 33:246–255.

JOHN, B., AND K. R. LEWIS. 1963. Chromosome marker. J. & A. Churchill, London.

KOSTOFF, D. 1939. Autosyndesis and structural hybridity in F_1 hybrid *Helianthus tuberosus* L. × *Helianthus annuus* L. and their consequences. Genetics 21:285–300.

PELLEW, C., AND F. M. DUNHAM. 1916. The genetic behaviour of the hybrid *Primula kewensis,* and of its allies. J. Genet. 5:159–182.

STEBBINS, G. L., JR. 1950. Variation and evolution in plants. Columbia University Press, New York.

TOBGY, H. A. 1943. A cytological study of *Crepis fuliginosa,*

C. neglecta, and their F_1 hybrids and its bearing on the mechanism of phylogenetic reduction in chromosome number. J. Genet. 45:67–111.

WINGE, O. 1917. The chromosomes: their numbers and general importance. Compt. Rend. Trav. Lab. Carlsberg 13:131–276.

SUPPLEMENTARY REFERENCES

BABCOCK, E. B. 1947. The genus *Crepis*. I. Univ. California Pub. Bot., vol. 21.

DOBZHANSKY, T. 1951. Genetics and the origin of species, 3d ed. Columbia University Press, New York.

KARPECHENKO, G. D. 1927. Polyploid hybrids of *Raphanus sativus* L. × *Brassica oleracea* L. Bull. Appl. Bot. Genet. Plant Breed. 17:305–410.

PATTERSON, J. T., AND W. S. STONE. 1952. Evolution in the genus *Drosophila*. Macmillan, New York.

WHITE, M. J. D. 1954. Animal cytology and evolution, 2d ed. Cambridge University Press, Cambridge.

———. 1969. Chromosomal rearrangements and speciation in animals. Ann. Rev. Genet. 3:75–98.

GLOSSARY

ACENTRIC. Applied to a chromosome or chromosomal segment lacking a centromere.

ACROCENTRIC. Applied to a chromosome whose centromere is very close to one end.

ALLOPOLYPLOID (Clausen, Keck, and Hiesey, 1945). Polyploid cell or individual with nonhomologous genomes.

AMPHIPLOID (Clausen, Keck, and Hiesey, 1945). An allopolyploid resulting from the doubling of the chromosomes in an interspecific hybrid (= amphidiploid).

ANAPHASE (Strasburger, 1884). Stage of mitosis or meiosis when daughter chromosomes or homologous chromosomes proceed toward opposite poles of the spindle apparatus.

ANEUPLOID (Täckholm, 1922). Cell, tissue, or individual with more or less than the standard somatic, gametophytic, or gametic chromosome number.

ANTHER. Part of stamen in which pollen grains are produced.

ASYNAPTIC (Beadle, 1931). Refers to the failure of homologous chromosomes to pair during the first meiotic division.

AUTOALLOPOLYPLOID (Kostoff, 1939). Cell or individual with chromosome sets showing the pairing characteristics of autopolyploidy and allopolyploidy. Autoalloploid species are generally hexaploids or higher levels of polyploidy.

Autopolyploid (Kihara and Ohno, 1926). Cell, tissue, or individual with three or more homologous sets of chromosomes.

Autosome (Montgomery, 1904). A chromosome not directly involved in sex determination.

Barr body. See *sex chromatin.*

Basic number (X). The chromosome number of the gametophyte or gametic nucleus in a diploid species of a genus with a polyploid series.

B chromosome (Randolph, 1928). Supernumerary chromosome; different from standard (A) chromosome in morphology, behavior, and genetic constitution.

Bivalent (Häcker, 1892). A pair of homologous chromosomes at diplotene, diakinesis, or metaphase I.

Centromere (Waldeyer, 1903). Region of the chromosome associated with fibers from the spindle apparatus during nuclear division.

Certation (Nilsson, 1915). The competition between pollen tubes of different genotypes, growing at different rates in the style, in effecting the fertilization of eggs.

Chiasma (Janssens, 1909). The site of the mutual switching of nonsister chromatids of homologous chromosomal segments; observed at diplotene, diakinesis, and metaphase I.

Chiasma interference (Mather, 1933). The more (negative) or less (positive) frequent occurrence of a second chiasma in the vicinity of the first chiasma than expected by chance.

Chiasma terminalization (Darlington, 1929). The shifting of a chiasma from its original site toward the end of the chromosome from earliest diplotene to the onset of anaphase I.

Chromatid (McClung, 1900). One of the two visible strands in the chromosome associated with one centromere.

Chromatid interference. See *interference.*

Chromatid tetrad. The four chromatids (two per chromosome) of a bivalent.

Chromocenter. Dark staining body representing a heterochromatic chromosomal segment in the resting nucleus; the pooled heterochromatic segments of the salivary-gland chromosomes of *Drosophila.*

Chromomere (Wilson, 1896). One of the darkly staining "beads" of different sizes and spacing along the length of the chromosome; readily observed during leptotene, zygotene, or early prophase.

CHROMOSOMAL INTERFERENCE. See *interference.*

CHROMOSOME (Waldeyer, 1888). Linear organelle within the resting nucleus of eukaryotes; usually detected by light microscopy during nuclear division.

CHROMOSOME COMPLEMENT (Darlington, 1932). The chromosomes of the gametic nucleus.

COMPENSATING TRISOMIC (Blakeslee, 1927). An individual with an extra chromosome ($2n + 1$) in which a missing standard chromosome is present in segments associated with a segment of nonhomologous chromosomes.

CROSSING-OVER (Morgan and Cottell, 1912). The event(s) responsible for the recombination of linked genes.

CYTOKINESIS (Whitman, 1887). Cell division.

DAUGHTER CHROMOSOMES. The chromosomes from chromatids previously associated with one centromere at metaphase I or metaphase II.

DEFICIENCY (Bridges, 1917). The loss of a chromosomal segment (= deletion).

DESYNAPSIS. The complete separation of homologous chromosomes at diplotene previously paired at pachytene.

DIAKINESIS (Häcker, 1897). The end of both the diplotene substage and prophase I stage of meiosis.

DICENTRIC. Applied to a chromatid or chromosome with two centromeres.

DIOECIOUS. Plant species in which different individuals have male or female flowers.

DIPLOID. Usually a somatic cell or individual with pairs of homologous chromosomes.

DIPLOTENE (von Winiwarter, 1900). Substage of prophase I when the homologous chromosomes first exhibit chiasmata to form bivalents.

DISOMIC (Blakeslee, 1921). Applied to a cell, tissue, or individual with pairs of homologous chromosomes.

DOUBLE FERTILIZATION. The fusion of one sperm nucleus with the egg nucleus and of the second sperm nucleus with the two polar nuclei of the embryo sac to produce the embryo and endosperm in the seed.

DUPLICATION (Bridges, 1919). An added chromosome or chromosomal segment.

DYAD. The two chromatids in a chromosome from diplotene to metaphase II.

DYSPLOIDY (Tischler, 1937). Different diploid numbers of species in a genus, other than a polyploid series.

EMBRYO SAC (Hofmeister, 1849). Female gametophyte in higher plant species.

ENDOSPERM. Polyploid tissue in the seed to provide nutrients to the developing embryo; one product of double fertilization.

EUCHROMATIN (Heitz, 1928). Chromatin in chromosomes or chromosomal segments with the usual cycle of coiling and normal stainability during nuclear division.

EUKARYOTE. Organism with chromosomes enclosed within a nuclear membrane and distributed to daughter cells by mitosis or meiosis.

EUPLOID (Täckholm, 1922). Cell, tissue, or individual with one or more homologous sets of chromosomes.

GAMETOPHYTE (Hofmeister, 1851). The haploid individual in plants originating from a spore and producing the gametic nucleus or nuclei.

GENOME (Winkler, 1920). The gametophytic or gametic chromosome set (X number) in a basic diploid species.

GENOMIC ALLOPOLYPLOID (Stebbins, 1947). Polyploid species with two or more nonhomologous genomes.

HALF CHROMATID. One of the two strands in a chromatid.

HAPLOID (Strasburger, 1905). Cell, tissue, or individual with the gametic rather than the somatic or sporophytic chromosome number.

HEMIZYGOUS. Applied to the presence of one rather than both alleles in a diploid because of a deficiency; genes on the X chromosome in an XY individual.

HETEROCHROMATIN (Heitz, 1929). A chromosomal segment or chromosome with an abnormal degree of contraction or stainability.

HOMOEOLOGOUS (Huskins, 1932). Applied to partially homologous chromosomes in different genomes.

HYPERPLOID (Bělăr, 1928). Cell, tissue, or individual with one or more added chromosomal segments or chromosomes.

HYPOPLOID. Cell, tissue, or individual deficient for one or more chromosomal segments or chromosomes.

INTERCALARY. Applied to the region between specified sites of the chromosome (= interstitial).

INTERCHANGE COMPLEX. Association of four or more chromosomes in a translocation heterozygote at diakinesis or metaphase I.

INTERFERENCE. CHROMOSOMAL (Muller, 1916): the occurrence of one crossing-over reduces the probability of a second crossing-over in its vicinity. CHROMATID (Mather, 1933): the nonrandom involvement of the chromatids in a double crossing-over.

INTERSEX (Goldschmidt, 1915). An individual of a species with different sexes displaying reproductive organs of each sex and with one genotype in all the somatic cells.

INVERSION (Sturtevant, 1926). A reversal of a segment within a chromosome.

ISOCHROMOSOME (Darlington, 1940). A chromosome with homologous arms.

KARYOKINESIS (Schleicher, 1878). Nuclear division.

KARYOTYPE. A pictorial or diagrammatic presentation of the metaphase chromosomes of the complement of an individual or a species.

LEPTOTENE (von Winiwarter, 1900). The earliest substage of prophase I when the chromosomes first appear and are not yet paired.

MEGASPORE. Product of meoisis from a megasporocyte.

MEGASPOROCYTE. A meiocyte yielding megaspores.

MEGASPOROGENESIS. The production of megaspores from megasporocytes by meiosis.

MEIOCYTE. A cell undergoing meiosis.

MEIOSIS (Farmer and Moore, 1905). The two nuclear divisions of a specialized cell (meiocyte) with a diploid number, yielding four haploid nuclei. The first nuclear division is assigned roman numeral I and the second, roman numeral II (Gregoire, 1904).

METACENTRIC. Applied to a chromosome whose centromere is equidistant from the ends.

METAPHASE (Strasburger, 1884). The stage during mitosis or meiosis when the centromeres of the chromosomes form an equatorial plane between the poles of the spindle apparatus.

MICROSPORE. Product of meiosis in a microsporocyte.

MICROSPOROCYTE. Meiocyte in the anther of higher plants yielding microspores (= pollen mother cell).

MICROSPOROGENESIS. The production of microspores from microsporocytes by meiosis.

MITOSIS (Flemming, 1882). The division of a nucleus so that the daughter nuclei have the same chromosome number as the parental nucleus (prophase, metaphase, anaphase, telophase).

MONOPLOID (Langlet, 1927). Cell, tissue, or individual with one genome.

MULTIVALENT. An association of three or more homologous or partially homologous chromosomes at diplotene, diakinesis, or metaphase I.

n, $2n$, $3n$, $4n$, . . . Symbols for levels of autoploidy (monoploidy, diploidy, autotriploidy, autotetraploidy, etc.).

NEOCENTRIC ACTIVITY. Association between a site on the chromosome other than the centromere and the fibers of the spindle apparatus.

NONDISJUNCTION. Distribution of sister chromosomes to the same pole at anaphase, anaphase I, or anaphase II of homologous chromosomes to the same pole.

NUCLEOLUS. A small spherical body within the nucleus and associated with a particular region in one or more chromosomes (nucleolus-organizing chromosomes).

NUCLEUS (Brown, 1831). A spherical body in the cell, staining deeply with basic dyes.

NULLISOMIC (Blakeslee, 1921). Cell or individual lacking one pair of chromosomes ($2n - 1$ bivalent).

PACHYTENE (von Winiwarter, 1900). A substage of prophase I when the homologous chromosomes are paired from end to end.

PARACENTRIC. Applied to an inversion not including the centromere.

PERICENTRIC. Applied to an inversion including the centromere.

PISTIL. Organ of a higher plant consisting of the ovary, style, and stigma.

PLOIDY. The number of genomes in a cell or individual (monoploidy, diploidy, triploidy, etc.) without reference to the homology or nonhomology of the genomes.

POLLEN MOTHER CELL (PMC). Microsporocyte.

POLYNUCLEOTIDE. Linear sequence of nucleotides in DNA or RNA.

PROKARYOTE. Organism with chromosomes not enclosed within a nuclear membrane and not distributed to daughter cells by mitosis or meiosis.

PROPHASE (Strasburger, 1884). The first stage in somatic nuclear division.

PSEUDOISOCHROMOSOME (Caldecott and Smith, 1935). A translocation chromosome with most of one arm homologous to the other arm.

QUADRIVALENT. An association of four homologous chromosomes at diplotene, diakinesis, or metaphase I.

RENNER COMPLEX. A group of chromosomes or of phenotypes distributed as a unit during meiosis (*Oenothera*).

SATELLITE (Navachin, 1912). A segment "separated" from the remainder of the chromosome by a constriction.

SECONDARY TRISOMIC. Cell or individual with an extra chromosome (isochromosome) whose arms are homologous.

SEGMENTAL ALLOPOLYPLOID (Stebbins, 1947). An allopolyploid with homologous chromosomal segments in different genomes.

SEMISTERILITY (Belling, 1914). The abortion of approximately 50 percent of the male and female gametophytes from a heterozygous reciprocal translocation in higher plants.

SEX CHROMATIN (Barr and Bertram, 1949). Very small, darkly staining body in the resting nucleus of mammalian cells, indicating an inactivated X chromosome (= Barr body).

SPORE. Product of microsporogenesis or megasporogenesis.

SPOROPHYTE. The diploid, asexual individual in plants.

STAMEN. Male organ of a plant, including the filament and anther, containing pollen grains.

STIGMA. The part of the pistil of the flower on which the pollen grain germinates and produces the pollen tube.

STRAND. A visible linear structure (chromatid) in the chromosome; polynucleotide in the DNA double helix.

STYLE. The "stalk" with the stigma at its apex.

SUBMETACENTRIC. Applied to a chromosome whose centromere is not exactly equidistant from the ends.

SUPERNUMERARY CHROMOSOME. B chromosome.

SYNAPSIS. Pairing of chromosomes during prophase I.

SYNAPTINEMAL COMPLEX (Moses, 1958). A complex structure between paired chromosomes; detected by electron microscopy.

TELOCENTRIC (Darlington, 1939). Applied to a chromosome with a terminal centromere.

TELOMERE (Muller, 1940). The end of an unbroken chromosome.

TERMINALIZATION (Darlington, 1929). The reduction in the number of chiasmata from diplotene to metaphase I.

TETRAD (Nemec, 1910). The four chromatids of a bivalent.

TETRAPLOID. A cell, tissue, or individual with four genomes.

TETRASOMIC (Blakeslee, 1921). A cell, tissue, or individual with an extra pair of chromosomes ($2n + 1$ bivalent).

TRANSLOCATION. Transposition of one chromosomal segment (simple translocation) or mutual exchange of chromosome segments, not by crossing-over (reciprocal translocation).

TRIPLOID. A cell, tissue, or individual with three genomes.

TRISOMIC (Blakeslee, 1922). A cell, tissue, or individual with an extra chromosome ($2n + 1$).

TRIVALENT. An association of three, usually homologous, chromosomes.

UNIVALENT. A single, unpaired chromosome at the first meiotic division.

VARIEGATION. Tissues of different phenotypes in an individual, resulting from extrachromosomal segregation, mutation, or chromosomal aberration.

X, 2X, 3X, etc. Symbol for the number of genomes in the gametic nucleus of species in a genus with a polyploid series.

ZYGOTENE (Gregoire, 1907). The substage of prophase I in which the homologous segments of chromosomes initiate pairing.

INDEX